Reviews of Physiology, Biochemistry, and Pharmacology 145

Springer
Berlin
Heidelberg
New York
Barcelona
Hong Kong
London
Milan
Paris
Tokyo

Reviews of

145 Physiology Biochemistry and Pharmacology

Editors
S. G. Amara, Portland E. Bamberg, Frankfurt
M. P. Blaustein, Baltimore H. Grunicke, Innsbruck
R. Jahn, Göttingen W. J. Lederer, Baltimore
A. Miyajima, Tokyo H. Murer, Zürich
N. Pfanner, Freiburg G. Schultz, Berlin
M. Schweiger, Berlin

With 20 Figures and 9 Tables

Springer

ISSN 0303-4240
ISBN 3-540-43520-4 Springer-Verlag Berlin Heidelberg New York

Libary of Congress Catalog-Card Number 74-3674

Springer-Verlag Berlin Heidelberg New York
a member of BertelsmannSpringer Sciences & Business Media GmbH

http://www.springer.de

Production: PRO EDIT GmbH, 69126 Heidelberg
Printed on acid-free paper – SPIN: 10875588 27/3130Re-5 4 3 2 1 0

Contents

Selenium-Containing Proteins in Mammals and Other Forms of Life
By Antonios Kyriakopoulos and Dietrich Behne
(With 1 Figure und 2 Tables) 1

Constitutively Active and G-Prozein Coupled Inward Rectifier K^+ Channels: Kir2.0 and Kir 3.0
By Peter R. Stanfield, Coventry, UK, Shigehiro Nakajima, and Yasuko Nakajima, Chicago, Il, USA
(With 16 Figures and 5 Tables) 47

Chloroplast Quest: A Journey from the Cytosol Into the Chloroplast and Beyond
By Ute C. Vothknecht and Jürgen Soll, Munich, Germany
(With 3 Figures and 2 Tables). 181

Indexed in Current Contents

Selenium-Containing Proteins in Mammals and Other Forms of Life

A. Kyriakopoulos and D. Behne

Hahn-Meitner-Institut Berlin
Department Molecular Trace Element Research in the Life Sciences
Glienicker Str. 100, 14109 Berlin, Germany

Contents

1 Introduction 3
2 Milestones in Selenoprotein Research 4
3 General Aspects 5
  3.1 Selenium and Sulfur: Comparison of Relevant Properties 5
  3.2 Formation of Selenocysteine-Containing Proteins 6
4 Selenium-Containing Proteins in Various Forms of Life 7
  4.1 Selenium-Containing Proteins in Plants 7
  4.2 Selenium-Containing Proteins in Yeasts 9
  4.3 Selenium-Containing Proteins in Bacteria and Archaea 10
    4.3.1 Formate Dehydrogenases 10
    4.3.2 Reductases 12
    4.3.3 Hydrogenases 13
    4.3.4 Other Selenoproteins 14
    4.3.5 Selenium-Dependent Enzymes 15
5 Selenium-Containing Proteins in Mammals 16
  5.1 General Points 16
    5.1.1 Selenium and Selenium-Containing Proteins in Mammalian Tissues 16
    5.1.2 Chemical Forms of Selenium in the Mammalian Selenium-Containing Proteins 18
  5.2 Selenoenzymes 20
    5.2.1 Glutathione Peroxidases 20
    5.2.2 Iodothyronine Deiodinases 24
    5.2.3 Thioredoxin Reductases 26
    5.2.4 Selenophosphate Synthetase 2 28
  5.3 Selenoproteins with Functions still Unknown 28
    5.3.1 Selenoprotein P 28
    5.3.2 Selenoprotein W 29
    5.3.3 15 kDa selenoprotein 29

Reviews of Physiology, Biochemistry, and Pharmacology, Vol. 145

5.3.4 18 kDa selenoprotein 30
5.3.5 Further Selenoproteins Identified in Silico 30
5.4 Selenium-Containing Proteins not yet Identified 30
6 Conclusions and Outlook 32
References 34

1 Introduction

Although selenium was discovered by Berzelius as early as 1817, it did not attract the attention of the life sciences until the 1930s when poisoning to grazing animals was attributed to the ingestion of plants with high selenium concentrations (Franke and Potter 1935). With the development of analytical methods with sufficiently low limits of detection the element was then found to be present at very low concentrations in a variety of living organisms, but was thought to be taken up only accidentally and metabolized due to its chemical similarity to sulfur. In the middle of the last century, however, several investigations showed that selenium is required by prokaryotic as well as eukaryotic cells.

Initial information in this field came in 1954 from the results of a study on bacteria, which showed that selenite is needed for stimulation of the enzymatic dehydrogenase activity in strains of *Escherichia coli* (Pinsent 1954). In 1957 it was found that liver necrosis, induced in rats by feeding a purified, vitamin E-deficient diet, could be prevented by the addition of selenium (Schwarz and Foltz 1957). In the same year it was also reported that selenium was effective in preventing exudative diathesis in chicken (Patterson et al. 1957). This vitamin E-responsive condition is characterized by leakage of plasma into subcutaneous spaces of the abdomen and breast. Subsequent investigations showed that selenium is an essential element for animals and that selenium deficiency can lead to various disorders (Combs and Combs 1986b). Interest in the element was further increased when it was shown that selenium has an protective effect against Keshan disease, an endemic cardiomyopathy which occurred in selenium-deficient areas of China (Keshan Disease Research Group 1979).

With the exception of some selenium-accumulating plants the organisms contain the main part of selenium in the form of proteins. The investigation of the formation, metabolism and functions of the selenium-containing proteins has therefore been a major field in selenium research. As the present studies in this area are concerned above all with the selenoproteins in the mammalian organisms and their role in health and disease, this review will deal mainly with the mammalian selenium compounds and will only summarize the findings on the selenium-containing proteins in some other forms of life such as bacteria, yeasts and plants.

2 Milestones in Selenoprotein Research

Ever since the discovery of the beneficial effects of selenium in bacterial (Pinsent 1954) and mammalian cells (Schwarz and Foltz 1957) the question of the mechanisms by which selenium exerts its biological functions have been of great interest. First explanations were found in 1973 when links between selenium and enzymatic reactions were established. In mammals, glutathione peroxidase was identified as a selenium-containing enzyme (Rotruck et al. 1973; Flohé et al. 1973), and in bacterial organisms selenium was shown to be a component of glycine reductase (Turner and Stadtman 1973) and of a formate dehydrogenase (Andreesen and Ljungdahl 1973). In both glycine reductase (Cone et al. 1976) and glutathione peroxidase (Forstrom et al. 1978; Wendel et al. 1978) selenium was found to be present in the form of selenocysteine. Based on the complete amino acid sequence of glutathione peroxidase isolated from bovine erythrocytes (Günzler et al. 1984), the DNA analysis of the mouse glutathione peroxidase gene revealed that a TGA codon, which normally acts as termination codon, is responsible for the insertion of selenocysteine into the amino acid chain of this mammalian enzyme (Chambers et al. 1986). The investigation of a formate dehydrogenase from E. coli showed that in bacteria, too, selenocysteine is encoded by the TGA codon (Zinoni et al. 1986). Continuing from these findings a series of studies on the pathways of selenocysteine incorporation was carried out which resulted in the elucidation of the different steps of the biosynthesis of bacterial selenoproteins and the identification of selenocysteine as the 21st encoded amino acid (Böck et al. 1991a,b). Studies on the biosynthesis of eukaryotic selenoproteins revealed the same basic principles but also some considerable differences (Berry et al. 1993; Low and Berry 1996).

Parallel to the deciphering of the mechanisms of selenoprotein formation, research has been concerned with the identification of novel selenoproteins and their biological functions. By labeling rats and mice in vivo with ^{75}Se numerous selenium-containing proteins have been found to exist in the mammalian tissues. The discovery of a hierarchy in selenium distribution which ensures preferential supply of the element to certain selenium compounds in certain tissues, with the glutathione peroxidase being last in this ranking order (Behne et al. 1988), was an indication of the biological significance of these compounds. A second important physiological function in addition to that of the different glutathione peroxidases was detected when selenium was identified as a catalytically active component of the type I iodothyronine deiodinase (Behne et al. 1990a; Arthur et al. 1990), a relationship which opened up a new field of research into the roles of selenium in the iodothyronine deiodinases and in thyroid hormone metabolism. This was followed by the findings that thioredoxin reductase (Tamura and Stadtman 1996), a well-known enzyme with key roles in various metabolic processes and selenophosphate synthetase 2 (Guimaraes et al. 1996), which is needed in the biosynthesis of selenocysteine, likewise contain selenocysteine in their active centers. Recent studies have shown that the essentiality of selenium is not only due to the functions of the selenoenzymes in the physiological processes in a multitude of mammalian tissues but that there are also very specific roles in spermatogenesis where the phospholipid hydroperoxide glutathione peroxidase (Ursini et al. 1999) and the sperm nuclei glutathione peroxidase (Pfeifer et al. 2001) have highly specialized tasks in certain spermatozoal compartments and are thus indispensable for male fertility. Valuable information on the existence of further mammalian selenoproteins was also obtained by scanning the nucleotide sequence databases for the elements necessary for decoding UGA as selenocysteine (Kryukov et al. 199; Lescure et al. 1999) and in the case of novel prokaryotic selenoproteins by searching the total genomic sequence of the archaeon *Methanococcus jannaschii* for open reading frames with internal UGA codons (Wilting et al. 1997).

3 General Aspects

3.1 Selenium and Sulfur: Comparison of Relevant Properties

Selenium is classified in group VIA of the periodic table below sulfur and its chemical and physical properties are very similar to those of the latter. Accordingly, both elements are distributed in the same way, but the sulfur concentration in the earth's crust and thus in living matter exceeds that of selenium by three to four orders of magnitude. Like sulfur, selenium occurs in the oxidation states -2, 0, +2, +4 and +6 and is present in compounds analogous to those of sulfur such as the inorganic selenides, selenites and selenates and the various organic compounds in which the element is present mainly in the -2 oxidation state such as dimethyl selenide, trimethyl selenonium, selenocysteine, selenomethionine and Se-methylselenocysteine.

The two elements differ, however, with regard to the reduction of their oxyanions. In biological systems the selenium compounds tend to be more easily reduced than sulfur, which is thermodynamically stable in the +6 state. Most important for the significance of selenium in biological systems, however, is the lower pK_a value of the selenohydryl group of selenocysteine (pK_a 5.24) as compared with that of the sulfhydryl group of cysteine (pK_a 8.25) (Huber and Criddle 1967). Accordingly, the selenols of the selenocysteine-containing proteins are anionic at physiological pH, while the thiols in the cysteine-containing proteins are mainly protonated under these conditions. Replacement of selenocysteine in the active center of a selenoenzyme with cysteine therefore results in a drastic decrease in its catalytic activity as was shown, for instance, in the case of formate dehydrogenase H from E. coli (Axley et al. 1991), type I iodothyronine deiodinase (Berry et al. 1991a) and thioredoxin reductase (Gasdaska et al. 1999a; Lee et al. 2000)

3.2 Formation of Selenocysteine-Containing Proteins

With sulfur present in large amounts in living matter as compared with selenium and the similarity in the chemical behavior of the two elements, highly selective mechanisms are necessary to prevent substitution of selenocysteine by cysteine during protein formation. After it had been shown in an experiment on rat liver glutathione peroxidase that serine provides the carbon skeleton of selenocysteine (Sunde and Evenson 1987), most of the steps which contribute to the selective incorporation of selenocysteine into the selenoproteins have been elucidated in studies on the formate dehydrogenase H in E. coli mutants (Böck et al. 1991a,b).

Selenocysteine insertion was found to be directed by a UGA codon in the open reading frame of the mRNA and to require a specific mRNA stem-loop structure which in bacteria is situated immediately downstream of the UGA codon, as well as four unique gene products, SelA, SelB, SelC and SelD. The UGA codon is recognized by the anticodon of a specific $tRNA^{(Ser)Sec}$ (SelC) which is first loaded with serine (Leinfelder et al. 1988). Selenocysteine synthase (SelA) then catalyzes the replacement of the side-chain oxygen of serine by selenium and thus the conversion of seryl- $tRNA^{(Ser)Sec}$ into selenocysteyl-$tRNA^{(Ser)Sec}$ (Forchhammer and Böck 1991). The selenium donor molecule for this reaction is selenophosphate (Leinfelder et al. 1990; Veres et al. 1992) which is produced from selenide and ATP by means of selenophosphate synthetase (SelD) (Veres et al. 1994; Lacourciere 1999). Selenocysteyl-$tRNA^{(Ser)Sec}$, GTP and a unique elongation factor SelB, homologous to the elongation factor EF-Tu for all other amino acid-tRNAs, form a ternary complex which binds to the stem loop structure of the selenoprotein mRNA and mediates the incorporation of selenocysteine into the protein at the ribosome (Heider et al. 1992).

The same basic principles of selenium biosynthesis were found in eukaryotes, with UGA as the codon responsible for selenocysteine insertion (Chambers et al. 1986), a specific $tRNA^{(Ser)Sec}$ which is first loaded with serine and then transformed into selenocysteyl- $tRNA^{(Ser)Sec}$ (Lee et al. 1989) and the production of selenophosphate by means of selenophosphate synthetases (Low et al. 1995; Guimaraes et al. 1996). Of the latter selenophosphate synthetase 2 was identified as a selenocysteine-containing enzyme (Guimaraes et al. 1996). The process of selenocysteine insertion differs, however, considerably, with regard to the position of the mRNA stem loop structure, named SECIS (selenocysteine insertion sequence) element, which in eukaryotes is not situated adjacent to the UGA codon but in the 3' untranslated region (Berry et al. 1993) and in this way is more similar to that of the archaea (Wilting et al. 1997). A eukaryotic selenocysteyl-tRNA$^{(Ser)Sec}$-specific elongation factor analagous to SelB has recently been identified (Tujebajeva et al. 2000). Together with the mammalian SECIS binding protein 2 (Copeland et al. 2000), it forms a complex which, different from the bacterial mechanism, allows a single SECIS element to decode multiple UGA codons as selenocysteines.

4 Selenium-Containing Proteins in Various Forms of Life

After the detection of the essential functions of selenium in bacteria and mammals research on selenium-containing proteins and their functions has been extended to a large variety of living organisms. Of those, yeasts and plants have been included in this review.

4.1 Selenium-containing Proteins in Plants

Although the first investigations on selenium in the biosphere were carried out on plants, knowledge of selenium-containing proteins in these organisms and of their metabolism and functions is relatively scarce. In the 1930s it was discovered that some plants growing on seleniferous soils enrich selenium in large quantities and that grazing animals are poisoned after ingestion of such plants (Franke and Potter 1935). In these so-called accumulator plants, most of which belong to the genus *Astragalus* or *Neptunia*, the selenium concentrations reach levels of up to several thousand mg Se/ kg, while in the non-accumulating plants they are around 1 mg Se/kg and below (Brown and Shrift 1982). In the accumulator plants selenium was found to be predominantly present in the non-protein fraction (Peterson and Butler 1962), mainly in the form of Se-methylselenocysteine and selenocystathionine (Brown and Shrift 1982). Selenomethionine and selenocysteine have been identified in the proteins of non-accumulator plants (Brown and Shrift 1980; Sathe et al. 1992a,b). and also in marine algae (Bottino et al. 1984).

Because of the similarity in their chemical properties selenium follows the same metabolic pathways as sulfur and is incorporated non-specifically in selenium derivatives of the sulfur metabolites. With the discovery of the essentiality of selenium in bacteria and mammals, however, the question arose whether selenium is also essential in plants and whether there are specific selenoproteins with biological functions in these organisms.

Several studies indicated that algae require selenium. The element was found to promote growth in several diatom species (Price et al. 1987; Harrison et al. 1988) and morphological changes were observed as an effect of selenium deficiency in the marine diatom *Thalassiosira pseudonana* (Doucette et al. 1987). In this species a selenium-dependent glutathione peroxidase was found which catalyzes the reduction of hydrogen peroxide (Price and Harrison 1988). The addition of selenite to the nutrient medium induced glutathione peroxidase in the green algae, *Chlamydomonas reinhardtii* (Yokota et al. 1988). It was purified and found to be a tetrameric selenium-containing enzyme with four identical subunits of 17 kDa (Shigeoka et al. 1991). The fact that it cross-reacted with an antibody raised against glutathione peroxidase from bovine erythrocytes showed that it is very similar to the mammalian enzyme.

In higher plants, however, the results of studies on selenium requirement and the existence of biologically active selenoproteins are still ambiguous. The finding that selenium stimulated growth in an *Astragalus* species (Trelease and Trelease 1939) suggested an essential function of the element in accumulator plants. However, later it was shown that the phosphate concentration in the nutrient solution used in this experiment was in the toxic range and hindered plant growth and that the beneficial effect of selenium was due rather to phosphate detoxification than to a direct essential function of the element (Broyer et al. 1972). A specific pathway for selenocysteine has been found in a selenium-accumulating *Astragalus* species, but this does not lead to the formation of selenoproteins. On the contrary, it prevents the element from entering sulfur-containing proteins (Neuhierl and Böck 1996). In this detoxification process a selenocysteine methyltransferase catalyzes the transformation of selenocysteine into Se-methyl-selenocysteine. The predominant part of selenium contained in abundance in the accumulator plants is present in free selenoamino acids such as Se-methyl-selenocysteine and selenocystathionine (Brown and Shrift 1982). Very little is known to date about the selenium-containing proteins in these organisms. In the seeds of *Lecythis ollaria*, for example, which contained about 10,000 mg Se/kg, 10% of the selenium was found to be protein-bound (Hammel et al. 1996). By analyzing the protein fractions separated by SDS-polyacrylamide gel electrophoresis the element was localized in extremely selenium-rich proteins with molecular masses below 20 kDa. The binding forms of selenium in these compounds have, however, not been identified.

In non-accumulator plants no evidence of a selenium requirement has as yet been found. The addition of up to 25 μg/l of selenium in the form of selenite to purified nutrient solutions had no effect on the growth of alfalfa and subterranean clover (Broyer et al. 1966). However, carefully controlled studies investigating the effects of both selenite and selenate on plant development in various species have still to be carried out. Glutathione peroxidase activity was detected in cultured cells of several plants (Drotar et al. 1985), but no test was performed to establish whether the activity stemmed from a selenium-containing enzyme.

In hydroponically grown soybeans (*Glycine max L.*) selenium was found to be covalently bound in lectin and sulfur-rich protein (Sathe et al. 1992a,b). The analysis showed, however, that the element was present in the form of selenomethionine incorporated non-specifically in place of methionine. After the addition of radioactively labeled selenite to the seedlings of mung beans (*Vigna radiata L*) and protein separation by polyacrylamide gel electrophoresis under non-denaturing conditions, a major selenium-containing protein with an apparent molecular mass in the range of 60–65 kDa was found in the mitochondrial fraction (Easwari and Lalitha 1995), but here again no further information on the binding form of the element and on the biological significance of this protein was presented. It is of interest in this respect, however, that in seedlings of this species grown with selenate, selenocysteine was found to be present in the proteins (Brown and Shrift 1980).

Although from the studies carried out so far no solid evidence for a selenium requirement has been obtained and no specific selenoproteins have as yet been identified, there is nevertheless some indication that the higher plants, too, have the genetic components necessary for selenoprotein biosynthesis. After the addition of radioactive selenite to cell cultures, ^{75}Se-labeled tRNA was found in a variety of plants such as wild carrots, tobacco, bamboo, rice, mung beans and soybean (Wen et al. 1988). These results suggested that selenium-containing tRNA might be universal in the plant kingdom. The fact that UGA-decoding Se-cysteyl-tRNAs have been identified not only in the diatom *Thalassiosira pseudonana* (Hatfield et al. 1991) but also in sugar beet (Hatfield et al. 1992) indicates that, as with bacteria and mammals, UGA can serve as a codon for selenocysteine insertion into polypeptides in higher plants, and this should therefore encourage research into specific selenoproteins in these organisms.

4.2 Selenium-Containing Proteins in Yeasts

Although the composition of yeast cells has been investigated in detail, little is known about the existence of specific selenium-containing proteins in these organisms. *Saccharomyces cerevisiae* cells grown in selenium-rich media have been used widely as a dietary selenium supplement. In these preparations more than 50 % of the element is present in the form of selenomethionine. The fact that after SDS-PAGE of ^{75}Se-labeled samples the tracer was found to be distributed over a large range of proteins suggested non-specific incorporation of the selenomethionine in place of methionine (Korhola et al. 1986). After the cultivation of *Saccharomyces cerevisiae* cells in a medium with only very small amounts of ^{75}Se-labeled selenite added and separation of the proteins in the subcellular fractions by means of SDS-PAGE, a cytosolic selenium-containing 25 kDa-band was found, which migrated with the same velocity as the subunit of glutathione peroxidase isolated from rat liver (Kyriakopoulos et al. 1998). Glutathione peroxidase activity was measured in a wild type strain of *Saccharomyces cerevisiae* (Galiazzo et al. 1987) and later in eleven species of yeast (Casalone et al. 1988). However, in both studies no test had been performed to find out whether the activity stemmed from a selenoenzyme. A cloned gene of *Schizosaccharomyces pombe*, homologous to the glutathione peroxidase genes, contained a cysteine codon in the UGA codon position of the mammalian gene (Yamada et al. 1999).

A selenocysteine-containing enzyme with glutathione peroxidase activity has, however, been located in *Saccharomyces uvarum* (Haas and Velten 1992). After growth of the cells in media supplemented with sodium selenite, subcellular fractionation, chromatographic and electrophoretic protein separation and selenium analysis of the fractions, three selenium-containing proteins were found. Of those, a mitochondrial 73 kDa-protein and a 83 kDa- protein present in the mitochondrial and cytosolic fractions were investigated in more detail. Both contained selenium in the form of selenocysteine. The 73 kDa-selenoprotein, which in SDS-PAGE had a selenium-containing 30 kDa-subunit, showed glutathione peroxidase activity. It reached maximum concentration at relatively low selenite levels in the medium, By contrast, the concentration of the 83 kDa-protein increased with rising levels of sodium selenite. This compound, which was identified as a glycoprotein, disintegrated into a selenium-containing 36,5 kDa-subunit. It was hypothesized that the 83 kDa-protein might be the inactive form of the 73 kDa-selenoenzyme and might serve as a store for this protein in the mitochondria.

4.3 Selenium-Containing Proteins in Bacteria and Archaea

Since the early discovery that selenium is needed for the enzymatic oxidation of formate in *Escherichia coli* (Pinsent 1954), several selenium-dependent enzymes have been identified in prokaryotes. In most of them the element is present in the form of genetically encoded selenocysteine. In addition, however, some enzymes were shown to be activated by selenium and were found to contain the element in a non-selenocysteine form. The selenium-containing proteins identified in bacteria and archaea are listed in Table 1.

4.3.1 Formate Dehydrogenases

Selenium-containing formate dehydrogenases were detected in several species of Enterobacteria (Enoch and Lester 1975, Cox et al. 1981, Heider et al. 1991) and Clostridia (Andreesen and Ljungdahl 1973, Leonhardt and Andreesen 1977, Wagner and Andreesen 1977) and two *Methanococcus* species (Jones and Stadtman 1981, Wilting et al. 1997). In addition to selenium these enzymes also contain a molybdenum or tungsten cofactor and iron-sulfur clusters.

Three selenocysteine-containing formate dehydrogenases could be distinguished in *Escherichia coli*. Under anaerobic conditions the formate dehydrogenase H (FDH_H) is synthesized, an 80 kDa-selenoprotein which is part of the

Table 1. Prokaryotic selenium-containing proteins

Protein	Organism	Significant studies
Bacteria		
Selenocysteine-containing proteins		
Formate dehydrogenase H	Escherichia coli	Cox et al. 1981
Formate dehydrogenase N	Escherichia coli	Enoch and Lester 1976
Formate dehydrogenase O	Escherichia coli	Sawers et al. 1991
Selenoprotein A of glycine reductase	Clostridium sticklandii	Cone et al. 1976
Selenoprotein B of glycine reductase	Eubacterium acidaminophilum	Wagner et al. 1999
Selenoprotein B of sarcosine reductase	Eubacterium acidaminophilum	Andreesen et al. 1999
Selenoprotein B of betaine reductase	Eubacterium acidaminophilum	Meyer et al. 1995
Proline reductase	Clostridium sticklandii	Kabisch et al. 1999
Seleno-Peroxiredoxin	Eubacterium acidaminophilum	Soehling et al. 2001
10 kDa Selenoprotein	Eubacterium acidaminophilum	Andreesen et al. 1999
[NiFeSe]-hydrogenase	Desulformicrobium baculatum	Voordrouw et al. 1989
Selenphosphate synthetase	Haemophilus influenza	Guimaraes et al. 1996
Selenium-containing proteins		
Xanthine dehydrogenase	Eubacterium barkeri	Schräder et al. 1999
Nicotinic acid hydroxylase	Clostridium barkeri	Gladyshev et al. 1996
Purine hydroxylase	Clostridium purinolyticum	Self and Stadtman 2000
CO dehydrogenase	Oligotropha carboxidovorans	Dobbek et al 1999
Archaea		
Selenocysteine-containing proteins		
Formate dehydrogenase	Methanococcus vannielii	Jones et al. 1979
Formylmethanofuran dehydrogenase	Methanopyrus kandleri	Vorholt et al. 1997
F_{420}-reducing hydrogenase	Methanococcus vannielii	Yamazaki 1982
F_{420}-non-reducing hydrogenase	Methanococcus voltae	Halboth and Klein 1992
F_{420}-non reducing hydrogenase	Methanococcus jannaschii	Wilting et al. 1997
Heterodisulfide reductase	Methanococcus jannaschii	Wilting et al. 1997
Selenophosphate synthetase	Methanococcus jannaschii	Guimaraes et al. 1996

formate-hydrogen lyase complex and catalyzes the conversion of formate into carbon dioxide and electrons which are transferred to a hydrogenase and thus reduce protons to elemental hydrogen (Peck and Gest 1957, Cox et al. 1981). Its crystal structure suggests a reaction mechanism that involves selenocysteine in proton abstraction and molybdenum and the Fe_4S_4 cluster in electron transfer (Boyington et al. 1997).

A second selenocysteine-containing formate dehydrogenase N (FDH_N) is synthesized when nitrate is present during anaerobic growth. It is associated with nitrate reductase and couples formate oxidation to nitrate reduction. FDH_N is a membrane-bound 600 kDa protein complex with four 110 kDa selenocysteine-containing subunits (Enoch and Lester 1975). A third enzyme capable of catalyzing the oxidation of formate under aerobic conditions is the formate dehydrogenase O (FDH_O) which is synthesized aerobically as well as anaerobically when nitrate is present. It likewise has a 110 kDa selenocysteine-containing subunit with a similar structure to that of FDH_N but is encoded by a different gene (Sawers et al. 1991).

A formate dehydrogenase with a large molecular mass was found in the archaeon *Methanococcus vannielii* in the soluble protein fraction. By means of amino acid analysis it was shown to contain selenium in the form of selenocysteine residues present in subunits of approximately 100 kDa (Jones et al. 1979, Jones and Stadtman 1981). By comparing ^{75}Se-labeled proteins with open reading frames containing internal UGA codons from the total genomic sequence a formate dehydrogenase with a calculated size of 77 kDa was identified in *Methanococcus jannaschii* (Wilting et al. 1997).

In several *Clostridia* species selenium-containing formate dehydrogenases are involved in the bidirectional conversion of carbon dioxide and formate. A formate dehydrogenase isolated from *Clostridia thermoaceticum* was shown to use NADPH as electron donor for the reduction of carbon dioxide. It has a molecular mass of about 340 kDa and a selenium-containing subunit of 96 subunits in which the element is most probably present in the form of selenocysteine and contains tungsten as a cofactor (Yamamoto et al 1983). Tungsten was also found to be a cofactor in the formate dehydrogenases found in *C. acidiurici* (Wagner and Andreesen 1977) *and C. formicoaceticum* (Leonhardt and Andreesen 1977) while in *C. cylindrosporum* molybdenum was required for maximum formate dehydrogenase levels (Wagner and Andreesen 1977). Detailed information on their composition and on the role of the selenium residue in the enzymatic functions is, however, not yet available.

4.3.2 Reductases

Glycine reductase is present in several anaerobic bacteria. It is involved in the process of energy conservation in the fermentative degradation of amino acids and purines, in which the electrons generated by the fermentation are used in the reductive deamination of glycine to acetylphosphate and ammonia (Arkowitz and Abeles 1989). The enzyme consists of three separate proteins PA, PB and PC (Turner and Stadtman 1973, Tanaka and Stadtman 1979). PA is an acidic 18 kDa glycoprotein which contains a selenocysteine residue (Turner and Stadtman 1973, Cone et al. 1976). The finding that glycine forms a covalent intermediate with selenocysteine in PA indicated a catalytic role for the selenocysteine residue (Arkowitz and Abeles 1990). Of the other two components, PB was likewise shown to contain a selenocysteine residue (Kreimer and Andreesen 1995). By the action of this selenol anion glycine bound to protein B is cleaved to give ammonia and a carboxymethylselenoether, which is transferred to protein A (Wagner et al. 1999).

Whereas most species known to contain glycine reductase are restricted to glycine as the electron acceptor, *Eubacterium acidaminophilum* can also grow on sarcosine (N-methylglycine) and betaine (N,N,N-trimethylglycine) as fermentation substrates. According to the substrates, specific substrate-binding proteins $PB_{Glycine}$, $PB_{Sarcosine}$ and $PB_{Betaine}$ are formed which were all shown to contain selenocysteine.(Wagner et al. 1999; Andreesen et al. 1999; Meyer et al. 1995). $PB_{Betaine}$ consists of two subunits of 48 and 45 kDa, with selenocysteine contained in the latter, while $PB_{Sarcosine}$. and $PB_{Sarcosine}$ consist of a 47 kDa selenocysteine-containing subunit and two further subunits of 25 and 22 kDa which stem from a posttranslationally processed proprotein (Andreesen and al. 1999).

D-proline reductase, which catalyzes the reduction of D-proline to 5-aminovalerate, is another reductase recently identified as a selenocysteine-containing protein (Kabisch et al. 19999. The enzyme, which was found in the cytoplasm *of Clostridium sticklandii*, has a molecular mass of about 870 kDa and consists of three subunits of 23, 26 and 45 kDa, with a selenocysteine residue in the 26 kDa protein B. Similar to the protein B of glycine reductase, the action of its selenol anion results in the cleavage of the carbon-nitrogen bond of the proline ring.

4.3.3 Hydrogenases

The system of hydrogenases is responsible for the oxidation of molecular hydrogen as well as for the reduction of protons to produce molecular hydrogen. In this system three classes of enzymes can be distinguished. [Fe]-hydrogenases which contain iron-sulfur clusters, [NiFe]-hydrogenases which contain nickel and iron-sulfur clusters and [NiFeSe]-hydrogenases which also contain selenium (Grahame 1988). [NiFeSe]-hydrogenases have been found in archaea growing on carbon dioxide and hydrogen as energy sources (Yamazaki 1982, Muth et al. 1987) and in sulfate-reducing bacteria (Teixeira et al. 1987; Fauque et al. 1988).

Two different types of [NiFeSe]-hydrogenases have been found in archaea. One type can reduce a specific deazaflavin cofactor F_{420},, the other only artificial acceptors. The first selenium-containing hydrogenase, a F_{420}-reducing enzyme, was detected in *Methanococcus vannielii* (Yamazaki 1982). It is a 340 kDa-enzyme with an $a_2b_4g_2$ subunit composition. The b subunit with a molecular mass of 42 kDa contains one selenocysteine residue. The molar ratio of selenium to nickel in this enzyme is 2 : 1 (Yamazaki 1987). The enzyme isolated from *M. voltae* likewise consisted of three subunits with molecular masses similar to that found in *M. vannielii* but contains nickel and selenium in equimolar amounts (Muth et al. 1987). Gene analysis indicated the existence of a further F_{420}-non-reducing [NiFeSe]-hydrogenase in this species (Halboth and Klein 1992). It consists of three subunits, with the selenocysteine residue contained in a posttranslationally processed, unusually small subunit of only 25 amino acids (Sorgenfrei et al. 1993). The three-dimensional structures of the active sites in the F_{420}-reducing and F_{420}-non-reducing hydrogenases are very similar and thus not influenced by the unusually short subunit of the latter enzyme (Sorgenfrei et al. 1997). These two hydrogenases and a further selenium-containing F_{420}-non-reducing hydrogenase were identified in *Methanococcus jannaschii* by comparing ^{75}Se-labeled proteins with open reading frames containing internal UGA codons from the total genomic sequence. In the case of the latter enzyme two UGA codons appear to code for a selenoprotein with two selenocysteine residues (Wilting et al. 1997).

Three [NiFeSe]-hydrogenase isoenzymes could be distinguished in the subcellular fractions of *Desulfovibrio baculatus* (now *Desulfomicrobium baculatum*)o which also have a 1 : 1 molar ratio of selenium to nickel. Like most of the non-selenium-containing hydrogenases they consist of two subunits with selenium present in the larger subunit with a molecular mass of 49, 54 and 62 kDa for the periplasmic, cytoplasmic and membrane-bound hydrogenase respectively (Teixeira et al. 1987). The presence of a TGA codon in the gene sequence indicated that these enzymes, too, contain selenium in the form of this amino acid (Voordouw et al. 1989). It has been shown that in the hydrogenase of *Dm. baculatum* selenium is directly coordinated to the active-site nickel atom (Eidsness et al. 1989; Garcia et al. 1999). Some differences in the catalytic properties (Fauque et al. 1988) and in the crystal structures (Garcia et al. 1999) of the [NiFeSe]-hydrogenases and [NiFe]-hydrogenases suggest that the two types of isoenzymes may be involved in different parts of the hydrogen metabolism and that the selenocysteine-containing enzymes may have a function in the production rather than in the uptake of hydrogen.

4.3.4 Other Selenoproteins

The genome of the archaeon *Methanopyrus kandleri* was found to harbor a gene which encodes the catalytic subunit of a tungsten formylmethanofuran dehydrogenase with an active site selenocysteine (Vorholt et al. 1997). This enzyme catalyzes the first step of methane formation by converting carbon dioxide to formylmethanofuran. In accordance with the identification of a selenocysteine residue the growth of *Methanopyrus kandleri* was stimulated by selenite. By comparing ^{75}Se-labeled proteins *in Methanococcus jannaschii* with open reading frames containing internal UGA codons from its total genomic sequence a further archaeal selenoprotein was identified as heterodisulfide reductase (Wilting et al. 1997). Heterodisulfide reductase is an iron-sulfur protein that catalyzes the reduction of the heterodisulfide product of the methanogenic thiol coenzymes to the free thiols.

By labeling *Eubacterium acidaminophilum* with ^{75}Se-selenite, two labeled proteins with molecular masses of 10 and 22 kDa were found which both were identified as selenoproteins. Peptide mapping of the 10 kDa selenoprotein showed that the motif involving selenocysteine resembled that of thioredoxin reductase from E. coli, but otherwise no similarities to known proteins had been found (Andreesen et al. 1999). By means of amplification and gene analysis the 22 kDa protein was identified as a selenocysteine-containing member of the peroxiredoxin family (Soehling et al. 2001). Peroxiredoxins catalyze the reduction of peroxides in a similar way to the selenocysteine-containing glutathione peroxidases, but in eukaryotic organisms they are cysteine-containing homologs. Thus, with this finding a further role of selenocysteine in the reductive destruction of peroxides has been shown to exist. Identification of the gene homologs in the genomes showed that selenophosphate synthetase is present in a selenocysteine-containing form not only in mammalian cells (Guimaraes et al. 1996) but also in the bacteria *Haemophilus influenzae* (Guimaraes et al. 1996) and *Eubacterium acidaminophilum* (Andreesen et al. 1999) and in the archaeon *Methanococcus jannaschii* (Guimaraes et al. 1996; Wilting et al. 1997). However, the fact that the selenocysteine-containing selenophosphate synthetase from *H. influenzae* did not exhibit a higher catalytic activity than the cysteine-containing enzyme from E. coli suggest a role of selenocysteine in the selenophosphate synthetase in *H. influenzae* that is not catalytic (Lacourciere and Stadtman 1999).

4.3.5 Selenium-Dependent Enzymes

Selenium was found to stimulate the enzymatic activity of xanthine dehydrogenase in *Clostridium acidiurici* and *Clostridium cylindrosporum* (Wagner and Andreesen 1979) and of nicotinic acid hydroxylase *in Clostridium barkeri* (Imhoff and Andreesen 1979). Both enzymes are complex molybdenum-iron-sulfur flavoproteins. The fact that selenium is lost during denaturation (Dilworth 1982; Dürre and Andreesen 1983) showed that it is not covalently bound as selenocysteine. Xanthine dehydrogenase from *Eubacterium barkeri* has a mass of 530 kDa and three types of subunits with molecular masses of

17,5 30 and 81 kDa. It contains molybdopterin as the molybdenum-complexing cofactor and 1 mol of selenium in a non-selenocysteine form per mol of native enzyme (Schräder et al. 1999). In nicotinic acid hydroxylase from *Clostridium barkeri* molybdenum is bound to molybdopterin and is coordinated with selenium which is essential for the enzymatic activity and which during denaturation is released as a low molecular weight compound (Gladyshev et al. 1996a). A selenium-containing purine hydroxylase has recently been purified from *Clostridium purinolyticum*. The loss of catalytic activity after the release of selenium indicates that here, too, the element is essential for purine hydroxylase activity (Self and Stadtman 2000). CO dehydrogenase from the aerobic bacterium *Oligotropha carboxidovorans*, which resembles xanthine dehydrogenase in many ways, likewise contains molybdenum and selenium in its active site, with the latter being attached to a cysteine residue as an S-selanylcysteine (Dobbek et al. 1999).

5. Selenium-Containing Proteins in Mammals

5.1 General Points

5.1.1 Selenium and Selenium-Containing Proteins in Mammalian Tissues

Selenium is very unevenly distributed among the different tissues. This was shown, for example, by analysis of the element in the different tissues of rats with a normal selenium status (Behne and Wolters 1983). The highest tissue selenium levels of about 6 mg/kg dry mass in testis and in kidney differed from the lowest in skeletal muscle and in brain by about one order of magnitude. The selenium content of most of the other tissues was in the range of 1 to 2 mg/kg dry mass. Sex-specific differences were found in the liver, with considerably higher values in the females. The enrichment of selenium in the male gonads ensures the incorporation of sufficient amounts of the element into the spermatozoa, which with a concentration of more than 20 mg/kg dry mass by far exceeded the selenium levels in the other compartments (Behne et al. 1986).

There are, however, no relationships between the tissue concentrations in the normal state and the pathological changes in selenium deficiency as these were found in some of the low- level tissues but also in the high-level compartments of the male reproductive system. Rats fed selenium-deficient diets over long periods of time produced spermatozoa with impaired motility and morphological anomalies which most frequently occurred in the midpiece region (McCoy and Weswig 1969; Wu et al. 1969). Severe selenium depletion in rats led to testicular atrophy and complete disruption of spermatogenesis (Behne et al. 1996a). Other animal tissues which in selenium deficiency or a combined low selenium and low vitamin E status were found to be affected include erythrocytes, eye, heart muscle, kidney, liver, pancreas, skeletal muscle, skin and smooth muscle (Combs and Combs 1986b). In humans selenium was shown to have a protective effect against Keshan disease, an endemic cardiomyopathy which occurred in selenium-deficient areas in China (Keshan Disease Research Group 1979). Selenium deficiency also seems to be a pathogenic factor in Kaschin-Beck disease, an osteoarthropathy which is endemic in selenium-deficient regions in Northeast Asia (Li et al. 1986). Cardiomyopathy (Volk and Cutliff 1986) and muscular disorders (Kopf et al. 1990) have been observed in patients on parenteral nutrition with a very low selenium intake.

One of the reasons for the tissue-specific differences in the susceptibility to selenium deficiency is the existence of a hierarchy in the selenium distribution in the mammalian organism (Behne and Höfer-Bosse 1984, Behne et al. 1988). With insufficient selenium intake regulation mechanisms strive to maintain the selenium levels in certain tissues such as the brain, the endocrine and the reproductive organs, and within the tissues the levels of certain selenoproteins. This is achieved by the preferential supply of these tissues with the element from the amounts taken up with the diet, but also by a redistribution of the metabolized element which in deficiency is excreted to a much lesser extent and is instead transported back to the priority target sites. Recent studies on rats showed that extreme selenium depletion over six generations resulted in a drastic decrease in the selenium concentrations in liver, skeletal muscle and blood below 1 % of the normal values and below 3 % in several other tissues such as heart, lung, prostate and stomach. In testis, too, the protective mechanisms, which work during the first stages of deficiency, had broken down. The brain, however, still contained 60 % of the concentration in the control animals, followed by spinal marrow, pituitary, thyroid, ovaries and adrenals. With these organs it seems to be virtually impossible to produce a state of severe selenium deficiency, even during extreme experimental depletion. They are thus well protected against losses during periods of insufficient selenium supply (Behne et al. 2000).

The puzzling finding of the first studies on the essential effects of selenium was the relationship between the element and vitamin E. Selenium deficiency exacerbates several tissue lesions induced by vitamin E deficiency and vice versa and several effects of the combined deficiency can be prevented by either the element or the vitamin (Combs and Combs 1986b). An explanation for this synergism was found when in 1973 selenium was identified as a component of glutathione peroxidase (Rotruck et al. 1973; Flohé et al. 1973). This enzyme catalyzes the reduction of peroxides and thus complements the effects of vitamin E in the cellular antioxidant defense system.

The antioxidant role of selenium in the form of glutathione peroxidase could explain several, but not all the effects of selenium deficiency. The suggestion that there might be further biologically active forms of selenium was supported by the finding that about two thirds of selenium present in the organism is not bound to this enzyme but is contained in other compounds (Behne and Wolters 1983).

Information on the presence of a larger number of selenium-containing proteins was then obtained in experiments in which rats and mice were labeled in vivo by administration of ^{75}Se-selenite and the selenium-containing proteins identified from the tracer distribution after chromatographic or gel electrophoretic separation (Hawkes et al. 1985; Danielson and Medina 1986; Calvin et al. 1987; Behne et al. 1988; Evenson and Sunde 1988). These compounds differed in their distribution among tissues (Behne et al. 1988) and subcellular fractions (Behne et al. 1990b; Kyriakopoulos et al. 1993; Kyriakopoulos et al. 1997; Kyriakopoulos et al. 2001). This suggested that they are part of several metabolic pathways of the element and might be involved in different intracellular processes. An interesting feature in selenoprotein research was the discovery that there is not only a tissue-specific regulation of the element but also an intracellular hierarchy, which, with insufficient selenium intake, results in the preferential incorporation of the element into certain selenium-containing proteins (Behne et al. 1988). Within most of the tissues the phospholipid hydroperoxide glutathione peroxidase and an 18 kDa selenoprotein were most preferentially supplied with the element, while the cellular glutathione peroxidase and the plasma glutathione peroxidase were last in this scale (D Behne & A Kyriakopoulos, submitted). The intracellular hierarchy in the expression of the selenoproteins may be regulated mainly by differences in the stability of their mRNAs in selenium deficiency (Bermano et al. 1996; Christensen and Burgener 1992; Sunde et al. 1997). The maintenance of the concentrations of certain high priority selenoproteins in periods of insufficient selenium supply may explain the fact that the total disruption of selenoprotein synthesis by knocking out the selenocysteyl-tRNA gene in mice resulted in early embryonal lethality (Bösl et al. 1997) while in rats fed a selenium-deficient diet for 16 generations no increased mortality could be observed (D Behne and A. Kyriakopoulos, unpublished data).

5.1.2
Chemical Forms of Selenium in the Mammalian Selenium-Containing Proteins

Selenium may be ingested in different chemical forms, In natural foods it is mainly present in the form of selenoamino acids such as selenocysteine and selenomethionine (Combs and Combs 1986a). It may enter the organism also as an inorganic compound such as selenite or selenate. In the organism selenium is either incorporated non-specifically into a large number of proteins, attached to specific selenium-binding proteins or incorporated into specific proteins which contain the element in form of genetically encoded selenocysteine-containing proteins and which have been defined as selenoproteins. In addition there are those proteins in which selenium has been detected but for which no information on its binding form is as yet available.

Specific selenoproteins: After ingestion of normal amounts of selenite, selenate or selenocysteine nearly all of the element is transported via an intermediary pool into specific selenocysteine-containing selenoproteins which are responsible for its biological effects. Their levels are homeostatically controlled and cannot be increased by additional selenium supplementation.

Selenium-binding proteins: There are also specific proteins in which selenium is only attached to the molecules. So far they include two proteins with molecular masses of 14 kDa (Bansal et al. 1989) and 56 kDa (Bansal et al. 1990) which were detected in mouse liver. The chemical form of selenium in these compounds is not known, but from the findings that the TGA codon responsible for selenocysteine incorporation is not present in the coding regions of the genes and that the levels of the two proteins are not dependent on dietary selenium supply, it can be concluded that the element is only firmly bound to these compounds. No information is as yet available on the function of the 56 kDa protein, but with the 14 kDa protein it has been suggested that it may act as a growth regulatory molecule and that by modulating its function selenium may inhibit cell growth.

In mouse tissues a protein with a molecular mass of about 17 kDa has been found which specifically binds selenite administered either in vitro or in vivo (Sani et al. 1988). Its function is not known but it has been suggested that it may be active in the intracellular transport of the element.

Non-specific selenium-containing proteins: In several distribution studies, of which only a few are mentioned here as examples, the retention of selenium in the tissues was found to be much higher when it was given as selenomethionine than as selenocystine, selenite or selenate (Deagen et al. 1987; Salbe and Levander 1990; Whanger and Butler 1988). In the investigation of the selenium-containing proteins in rat tissues after labeling with normal and large doses of selenite and selenomethionine it was shown that the higher tissue selenium contents were caused by incorporation into a large number of proteins (Behne et al. 1991). This is due to the fact that in the case of dietary selenomethionine only a part of the element is metabolized in the same way as the other selenium compounds. A certain percentage is deposited directly non-specifically into proteins in place of methionine and is therefore mainly found in methionine-rich proteins present in the organism in higher concentrations. This part that follows the metabolic pathways of methionine appears to be dependent only on the ratio of selenomethionine and methionine. It can therefore be influenced by changing the selenomethionine concentration in the diet as described above, but also by administering diets with low methionine levels (Waschulewski and Sunde 1988; Butler et al. 1989). In the latter case the increase in non-specific incorporation results in a decrease in the concentrations and effects of the specific selenoproteins (Waschulewski and Sunde 1988).

5.2 Selenoenzymes

All the mammalian selenoproteins identified to date are enzymes, with the selenocysteine residue responsible for their catalytic functions. The selenoenzymes known so far are listed in the top part of Table 2. They include the glutathione peroxidases, the iodothyronine deiodinases, the thioredoxin reductases and a selenophosphate synthetase. With the exception of the latter they are catalytically active in redox processes with thiols used as electron donors. Although their enzymatic functions have been established, with most of them information on their metabolic role and their biological significance is far from complete.

5.2.1 Glutathione Peroxidases

So far five selenocysteine-containing glutathione peroxidases have been detected: the cytosolic or classical glutathione peroxidase, a glutathione peroxidase found in the gastro-intestinal tract, the plasma glutathione peroxidase, the phospholipid hydroperoxide glutathione peroxidase and, as the most recent member of this family, another tissue-specific glutathione peroxidase which is only present in the sperm nuclei. The glutathione peroxidases catalyze the reduction of hydrogen peroxide and organic hydroperoxides and thus protect the cells from oxidative damage; Glutathione normally serves as the electron donor but there are cases where other thiols are oxidized in order to fulfill a specific biological role.

Cytosolic or Classical Glutathione Peroxidase

The cytosolic glutathione peroxidase (cGPx) detected in 1957 as an enzyme involved in the protection of erythrocytes against oxidative hemolysis (Mills 1957), was the first selenoprotein to be identified (Rotruck et al. 1973; Flohé et al. 1973). It is present in nearly all of the tissues but is unevenly distributed. In rats the measurement of glutathione peroxidase activity in the tissue cytosols, which almost completely stems from cGPx, showed the highest values in the liver and erythrocytes, while at the other end of the scale the levels in the skeletal muscle and the brain were lower by about two orders of magnitude (Behne and Wolters 1983). The enzyme, which consists of four identical selenocysteine-containing subunits with a molecular mass of about 22 kDa, catalyzes the reduction of hydrogen peroxide and various soluble organic peroxides. In this way it contributes to the antioxidant defense against reactive mo-

Table 2. Mammalia selenoproteins

Selenoprotein	Abbreviations used	Significant studies
With known functions		
Glutathione peroxidases	GPxs	
Cytosolic or classical GPx	cGPx, GPx1	Rotruck et al. 1973; Flohé et al. 1973
Gastrointestinal GPX	GI-GPx, GPx2	Chu et ak. 1993
Plasma GPx	pGPx GPx3	Takahashi et al. 1987
Phospholipid hydroperoxide GPx	PHGPx, GPx4	Ursini et al. 1985
Sperm nuclei GPx	snGPx	Pfeifer et al. 2001
Iodothyronine deiodinases		
Type 1 deiodinase	D1, 5'DI	Behne et al. 1990; Arthur et al. 1990
Type 2 deiodinase?	D2, 5'DII	
Type 3 deiodinase	D3, 5'DIII	Croteau et al. 1996
Thioredoxin reductases	TrxRs	
Thioredixub reductase 1	TrxR1	Tamura and Stadtman 1996
Thioredixub reductase 2	TrxR2	Gasdaska et al. 1999; Lee et al. 1999; Miranda-Vizuete et al. 1999; Watabe et al. 1999
Thioredixin reductase 3	TrxR3	Sun et al. 1999
Selenophosphate synthetase 2	SPS2	Guimaraes et al 1996
With unknown functions		
Selenoprotein P	SelP	Motsenbocker and Tappel 1982; Hill et al. 1991
Selenoprotein W	SelW	Vendeland et al. 1993; Vendeland et al. 1995
15 kDa selenoprotein	Sel15	Kalklösch et al. 1995; Gladyshev et al. 1998
18 kDa selenoprotein	Sel18	Kyriakopoulos et al. 1996
Selenoprotein R	SelR	Kryukov et al. 1999
Selenoprotein T	SelT	Kryukov et al. 1999
Selenoprotein N	SelN	Lescure et al. 1999
Selenoprotein X	SelX	Lescure et al. 1999
Selenoprotein Zf1	SelZf1	Lescure et al. 1999
Selenoprotein Zf2	SelZf2	Lescure et al. 1999

lecules and free radicals and complements the effects of vitamin E which acts as a free radical scavenger. However, the losses in cGPx activity in selenium-deficient animals did not lead to pathological changes (Sunde et al. 1997). Even after drastic selenium depletion in rats fed a selenium-deficient diet for sixteen generations, with a decrease in liver cGPx activity below the detection limit, no lesions were observed which could be attributed to the loss in cGPx activity (D Behne, unpublished data). The same was true with cGPx knockout mice which showed normal development (Ho et al. 1997). Effects were observed, however, after application of relatively high doses of paraquat in the knockout mice, which were affected to a much greater extent than the control animals (Cheng et al. 1998; de Haan et al. 1998). The fact that the mutation of a benign strain of coxsackie virus into a virulent myocarditis-inducing genotype, previously observed in selenium-deficient mice (Beck et al. 1994), also occurred in the knockout mice indicated that this mutation was due to the lack of cGPx (Beck et al. 1998). These effects together with the pathological changes observed in animals with combined selenium and vitamin E deficiency (Combs and Combs 1986b) suggest that under normal physiological conditions a low cGPx activity may be compensated for by other components of the antioxidative system but that the protective effects of cGPx are of particular importance when the organism is exposed to additional stress factors.

Gastrointestinal Glutathione Peroxidase

An indication for the existence of a further GPx was obtained from a cDNA sequence with more than 60 % identity to that of cGPx, detected in a human liver cDNA library (Akasaka et al. 1990). This gastrointestinal glutathione peroxidase (GI-GPx) is similar to cGPx in that it is a cytosolic selenoenzyme which consists of four identical selenocysteine-containing subunits with a molecular mass slightly below 22 kDa and catalyzes the reduction of various peroxides (Chu et al. 1993). Unlike cGPx, however, it is a tissue-specific enzyme which was found in the rat only in the gastro-intestinal tract. and in man only in this tract and in the liver. In the epithelium of the rodent gastrointestinal tract it contributes to about half of the total glutathione peroxidase activity (Esworthy et al. 1998). Studies in cell lines on its selenium-dependent expression showed that it is one of the high priority selenoproteins (Wingler et al. 1999). Due to its tissue specificity it has been suggested that GI-GPx may be a major component in the defense system against ingested lipid hydroperoxides (Esworthy et al. 1998) and may be of importance in the prevention of colon cancer (Chu et al. 1997).

Plasma Glutathione Peroxidase

Studies on the properties of the purified human plasma glutathione peroxidase (Takahashi et al. 1987; Broderick et al. 1987; Maddipati and Marnett 1987) and immunological tests with an antibody against this compound (Takahashi et al. 1987) showed the existence of a specific plasma glutathione peroxidase (pGPx). It was identified as a tetrameric GPx with a molecular mass of the subunits of approximately 23 kDa which differs from cGPx and GI-GPx in that it is a glycoprotein and is present in the extracellular fluids (Takahashi et al. 1987). It is expressed in various tissues from which it is secreted into the extracellular fluids, but the kidney has the highest concentration of pGPx mRNA and is the main site of production for this enzyme (Yoshimura et al. 1991; Chu et al. 1992; Avissar et al. 1994). In vitro experiments showed that like the other tetrameric glutathione peroxidases it catalyzes the reduction of hydrogen peroxides and various organic peroxides when glutathione is used as a substrate. However, its specific enzymatic activity is only 10 % of that of cGPx (Takahashi et al. 1987). Although it was identified more than ten years ago, its biological significance is still not clear. This is mainly due to the fact that the glutathione concentration in blood plasma is too low to serve as a suitable substrate for this enzyme. However, it has been shown that in this catalytic reaction thioredoxin and glutaredoxin are better suited as electron donors than glutathione (Björnstedt et al. 1994), and this finding might stimulate research into the biological role of pGPx.

Phospholipid Hydroperoxide Glutathione Peroxidase
A protein isolated from pig liver was found to exhibit GPx activity with phosphatidylcholine hydroperoxide, a substrate which did not react with cGPx (Ursini et al. 1982). Follow-up studies led to the detection of phospholipid hydroperoxide glutathione peroxidase (PHGPx) (Ursini et al. 1985). Unlike the three glutathione peroxidases described above it is a monomer with a molecular mass of 19.7 kDa. It is found in the tissues in both cytosolic and membrane-associated forms (Roveri et al. 1994). In vitro translation of the full-length PHGPx mRNA showed the existence of a PHGPx precursor with an additional N-terminal leader sequence which is responsible for the specific import of the enzyme into the mitochondria (Arai et al. 1996; Pushpa-Rekha 1995). Unlike the other tetrameric glutathione peroxidases PHGPx can directly reduce phospholipid and cholesterol hydroperoxides (Ursini et al. 1985; Thomas et al. 1990) and was thus considered primarily as a factor in the protective system against the oxidative destruction of biomembranes. In the meantime the results of numerous studies suggest that the enzyme may have important functions in the redox regulation of a variety of processes such as inflammation and apoptosis, although in most of these cases it is not yet known to what extent the other glutathione peroxidases may also be involved in these reactions. A significant role of PHGPx, only fulfilled by this enzyme, has, however, been found in spermatogenesis.

Earlier findings that selenium is highly enriched in the spermatozoa in the form of a selenoprotein located in the outer mitochondrial membrane (Calvin et al. 1981), that testicular selenium concentration rises sharply after the onset of puberty (Behne et al. 1986) and changes considerably due to gonadotropin-related interruption and the reconstitution of spermatogenesis (Behne et al. 1982) and that a 19.7 kDa selenoprotein, present not only in the testis but also in all the other tissues, is by far the most prominent selenium compound in the spermatozoa (Behne et al. 1988) were explained when it was shown that the sperm mitochondrial membrane selenoprotein is in fact PHGPx (Ursini et al. 1999) and that the increase in testis selenium during pubertal maturation is due to the abundant expression of this enzyme in the round spermatids (Maiorino 1998). However, while in the rat testis PHGPx activity is very high (Roveri et al. 1992), it is below the limit of detection in the epididymal spermatozoa (Ursini et al. 1999). Here the enzyme constitutes at least 50 % of the proteins present in the mitochondrial capsule but it is present in an inactive, oxidatively cross-linked form. These findings indicate a change in the biological role of PHGPx from that of an enzyme protecting against peroxide-induced damage to that of an inactive, but nevertheless important matrix component.

Sperm Nuclei Glutathione Peroxidase

Labeling of rats in vivo with ^{75}Se led to the detection of a 34 kDa selenoprotein which was present only in the testis and spermatozoa (Behne et al. 1988). It appeared after the onset of puberty and was localized in the nuclei of the late spermatids (Behne et al. 1997). During these stages of sperm development the nuclei undergo considerable changes characterized by the replacement of the histones by the protamines and the reorganization and condensation of the DNA, which result in compact, very tightly packed nuclei which are stabilized by cross-linking of the protamine thiols. The protein was identified as a specific sperm nuclei glutathione peroxidase (snGPx) which differs from PHGPx in its N-terminal sequence (Pfeifer et al. 2001). This sequence, which is encoded for by an alternative exon in the first intron of the PHGPx gene, is responsible for the specific biological role of snGPx. It contains a signal for the localization of the enzyme within the nuclei, where it is the only selenoprotein present, and a polyarginine-rich region by which it is attached to the DNA. In selenium-depleted rats where the concentration of snGPx had decreased to one third of the normal level, chromatin condensation was severely disturbed. It was possible to show that the enzyme acts as a protamine thiol peroxidase responsible for disulfide cross-linking and thus necessary for sperm maturation and male fertility.

5.2.2 Iodothyronine Deiodinases

The iodothyronine deiodinases catalyze the activation and inactivation of the thyroid hormones which regulate various metabolic processes and are indispensable for the normal development of the fetal brain. The family of the deiodinases consists of three members which differ with regard to their tissue distribution and their role in the deiodination of thyroxine and its metabolites. With the numerous reviews available on this subject only a few main findings will be presented.

Type 1 Deiodinase

A further important biological role of selenium in addition to that of the GPXs was detected in 1990. Two groups then showed that a membrane-bound selenium-containing 27 kDa protein, which had been found in the thyroid, liver and kidney of the rat (Behne et al. 1988), was identical to the subunit of type 1 deiodinase (D1). The protein was thus identified as a selenoenzyme (Behne et al. 1990a; Arthur et al. 1990). It was also shown that it contains one covalently bound selenium atom per molecule in its active center (Behne et al. 1990a). Cloning of the rat D1 cDNA proved the existence of an in-frame UGA codon responsible for the incorporation of selenocysteine into this enzyme (Berry et al. 1991b). D1 is located mainly in the thyroid, liver, kidney and pituitary. It can catalyze monodeiodination of the iodothyronines at the 5'-position of the phenolic ring or at the 5-position of the tyrosyl ring. 5'-deiodination of thyroxine (T_4) results in the formation of the biologically active hormone 3,3',5-triiodothyronine (T_3), whereas by means of 5-deiodination the inactive isomer 3,3',5' triiodothyronine (reverse T_3) is produced. 5-deiodination of T_3 and 5'-deiodination of reverse T_3 then lead to the inactive 3,3'-diiodothyronine (T_2). The biological role of D1 is to provide T_3 to the plasma, to inactivate T_4 and T_3 and to eliminate reverse T_3 from the circulation. The decrease in T_3 production from T_4 found in the liver of selenium-deficient rats (Behne et al. 1990a; Arthur et al. 1990) shows that an adequate selenium supply is important for thyroid hormone metabolism.

Type 2 Deiodinase

The type 2 deiodinase (D2) is membrane-bound, has subunits with molecular masses of about 30 kDa and is expressed predominantly in the brain, the brown adipose tissue, the pituitary and the placenta. It only catalyzes 5'-monodeiodination and converts T_4 to T_3 and reverse T_3 to T_2. Its main biological role is the local intracellular production of T_3 from circulating plasma T_4 in the tissues which express this enzyme and is thus a major factor in tissue-specific regulation. The question of whether D2 is a selenoprotein has not yet been completely settled. cDNA cloning has shown that D2 in amphibian tissue is a selenoenzyme (Davey et al. 1995) and evidence has been provided that the same is true of the mammalian D2 (Croteau et al. 1996; Salvatore et al. 1996). However, final proof is still not available, as in the clones assumed to code for selenocysteine-containing D2 in mammals the structures responsible for the incorporation of this amino acid have not yet been identified. A non-selenocysteine containing biologically active D2 subunit has recently been found (Leonard et al. 2000). Further studies are therefore needed to solve the question of the role of selenium in the enzymatic D2 activity.

Type 3 Deiodinase

Cloning of the type 3 deiodinase (D3) cDNA from the rat established that D3 is a selenoenzyme (Croteau et al. 1995). It is a 32 kDa selenoprotein which is mainly located in the central nervous system, the placenta, and the skin. D3 catalyzes the deiodination of the tyrosyl ring and is thus able to inactivate the thyroid hormones by producing reverse T_3 from T_4 and T_2 from T_3. In this way it protects the developing mammalian brain from exposure to excessive amounts of T_3 (Kaplan 1986) and regulates the supply of T_4 and T_3 from the mother to the fetus (Mortimer et al. 1996).

5.2.3 Thioredoxin Reductases

With the identification of a mammalian thioredoxin reductase as a selenoenzyme reductase (Tamura and Stadtman 1996) a third major physiological role of selenium has been detected. Mammalian thioredoxin reductases are a family of homodimeric flavoenzymes which are present in various tissues. In addition to the flavin and the active site of the prokaryotic homologs with their redox-active disulfide, they also contain selenocysteine as the penultimate C-terminal amino acid residue (Gladyshev et al. 1996b), which is indispensable for their enzymatic activity (Gromer et al. 1998). Thioredoxin reductases were named for their ability to catalyze the NADPH-dependent reduction of oxidized thioredoxin. Reduced thioredoxin is a central factor in cellular redox regulation. It provides reducing equivalents for various redox-dependent systems, e. g. for ribonucleotide reductase, essential for DNA synthesis and for the redox regulation of transcription factors, and has important functions in regulating cell growth and inhibiting apoptosis (Mustacich and Powis 2000). The significance of thioredoxin for the mammalian organism was shown in an experiment in which the disruption of the thioredoxin gene resulted in early embryonic lethality (Matsui et al. 1996). In addition to thioredoxin, mammalian thioredoxin reductases are able to use other substrates including hydroperoxides, dehydroascorbate and various enzymes and proteins (Holmgren and Björnstedt 1995). This broad substrate specificity has been attributed to the presence of selenocysteine situated in the flexible C-terminal extension (Gromer et al. 1998).

Thioredoxin Reductase 1

The first characterized mammalian selenium-containig protein of the thioredoxin reductases family was a selenium-containing 56 kDa protein, purified from ^{75}Se-labeled human lung cancer cells (Tamura and Stadtman 1996). This cytosolic enzyme, later named thioredoxin reductase 1 (TrxR1), is a dimer with two identical 56 kDa subunits and an isoelectric point of 5.2. The sequence of the TrxR1 cDNA, obtained from a human placental library, was found to have 44 % identity with that of the eukaryotic and prokaryotic glutathione reductases, but only 31 % with that of the prokaryotic thioredoxin reductases (Gasdaska et al. 1995). The human protein is predicted to have 524 amino acids including a conserved – Cys-Val-Asn-Val-Gly-Cys catalytic site and a selenocysteine containing C-terminal (Gasdaska et al. 1999)

Thioredoxin Reductase 2

A further mammalian selenocysteine-containing thioredoxin reductase was purified and characterized from mitochondria of several tissues. The enzyme is a first protein component in the mitochondrial thioredoxin-dependent reductase system. The mitochondrial thioredoxin reductase 2 (TrxR2), was described by four groups in 1999 either when its cDNA was cloned from human tissues (Gasdaska et al. 1999b), human adrenal (Miranda-Vizuete et al. 1999) and rat liver (Lee et al. 1999) or when its sequence was determined after purification of the protein from bovine adrenal cortex (Watabe et al. 1999). The molecular masses were around 56 kDa for the human and bovine proteins and about 53 kDa for the rat enzyme. The sequence identity between the mitochondrial TrxR2 and the cytosolic TrxR1 of the same species was found to be 54 % (Gasdaska et al. 1999b), 56 % (Miranda-Vizuete et al. 1999), 54 % (Lee et al. 1999) and 57 % (Watabe et al. 1999). TrxR2 differs from TrxR1 in that it has an N-terminal extension which was identified as a mitochondrial leader sequence (Miranda-Vizuete et al. 1999). The specific localization of TrxR2 in the mitochondria suggest that this enzyme provide a primary line of defense against H2O2 produced by the mitochondrial respiratory chain.

Further Thioredoxin Reductases

A thioredoxin reductase named here 3 (TrxR3), was purified from ^{75}Se-labeled mouse testis, where it is preferentially expressed (Sun et al. 1999). The deduced sequence of the human enzyme shows 70 % identity to that of TrxR1. It contains a long N-terminal extension and, with about 65 kDa, has a higher molecular mass than the other two isozymes.

It has recently been observed that TrxR1 isolated from mouse liver, mouse liver tumor and a human T-cell line exhibited considerable heterogeneity, which, as with the production of snGPx (Pfeifer et al. 2001), is due to alternative splicing of the first exons of the TrxR1 gene. By means of homology analyses, three isoforms of mouse and rat TrxR1 mRNA could be distinguished. Expression of multiple mRNA forms was also observed for human TrxR2 (Sun et al. 2001). By means of an algorithm which scans nucleotide sequence databases for mammalian selenocysteine insertion elements, two selenoproteins SelZf1 and SelZf2 were identified which share a common domain with TrxR2 and are probably produced by alternative splicing (Lescure et al. 1999). The findings of the two studies suggest the existence of further thioredoxin reductase species which may differ with regard to their distribution among tissues and subcellular compartments and may have specific biological roles. In the end can be said that all thioredoxin reductases contain a selenocysteine residue which is located at the carboxyl terminus and which is encoded by a UGA codon. In contrast with other members of the thioredoxin reductases (Bacteria, plant and yeast) are homodimers of 55-65 kDa.subunits Moroever they show a little homology to the bacterial, plant and yeast thioredoxin reductases, which are homodimers with the molecular mass of about 35 kDa per subunit (Gladyshev et al. 1999).

5.2.4 Selenophosphate Synthetase 2

Selenophosphate synthetase is indispensable for selenoprotein synthesis as it catalyzes the reaction of selenide with AMP. The product, selenophosphate, acts as the selenium donor for the biosynthesis of selenocysteine. In addition to selenophosphate synthetase 1, which contains threonine in its active center (Low et al. 1995), a selenocysteine-containing homolog with a molecular mass of about 50 kDa has been identified in various human and mouse tissues (Guimaraes et al. 1996). Information on the differences in the functions of the two enzymes in the biosynthesis of the mammalian selenoproteins is not yet available. The detection of a selenoenzyme which is involved in the production of the selenoproteins is of special interest with regard to the regulation of the mammalian selenium metabolism.

5.3 Selenoproteins with Functions still Unknown

5.3.1 Selenoprotein P

Following reports of the incorporation of ^{75}Se into a plasma protein in the rat (Millar 1972; Burk 1973) it was shown that this compound is not related to glutathione peroxidase (Herrman 1977) and that it contains selenium in the form of selenocysteine (Motsenbocker and Tappel 1982). Selenoprotein P (SelP) is a glycoprotein with a molecular mass of 43 kDa and constitutes more than 60 % of the plasma selenium (Read et al. 1990). Cloning of rat liver cDNA showed that SelP contains ten selenocysteines (Hill et al. 1991) and thus differs from all the other selenoproteins identified so far, which only have one selenocysteine residue per molecule or subunit. A cDNA obtained from bovine brain suggested the existence of a second SelP with 12 selenocysteine residues (Saijoh et al. 1995). SelP is mainly expressed in the liver but is also present in other tissues. Although it was the second selenoprotein to be detected, its function is still unknown. Because of its extracellular location and its high selenium content it was thought to act as a selenium transport protein (Motsenbocker and Tappel 1982). However, the fact that protection against diquat-induced liver lesions found after administration of selenium to selenium-deficient rats coincided with the appearance of SelP, suggested that it may act as an antioxidant (Burk et al. 1995). This hypothesis is supported by the finding that in human plasma it contributes to the destruction of peroxynitrite, thought to be an important factor in inflammatory toxicity (Arteel et al. 1996). Further studies are therefore needed to find out more about its significance and biological role.

5.3.2 Selenoprotein W

Selenoprotein W (SelW) was first purified from rat skeletal muscle and was shown to be a cytosolic selenoprotein with a molecular mass of slightly less than 10 kDa (Vendeland et al. 1993). Cloning of its cDNA indicated that it contains one selenocysteine residue per molecule (Vendeland et al. 1995). It is enriched in skeletal muscle and heart muscle, brain, testis and spleen, but was also found in a large number of other tissues (Sun et al. 1998). Its biochemical and physiological role is not yet known. However, the protein isolated from muscle tissue was found to contain glutathione which may have been bound as a reactant in an enzymatic redox cycle (Gu et al. 1999). The finding that glial cells with overexpressed levels of SelW were more resistant to peroxidation than normal cells (Sun et al. 2001) could likewise suggest a redox function.

5.3.3 15 kDa selenoprotein

Two 15 kDa selenoproteins with similar characteristics have been found in rat tissues (Kalcklösch et al. 1995) and in human T cells (Gladyshev et al. 1998). The rat 15 kDa selenocysteine-containing protein was identified by labeling the animals with ^{75}Se, separation of the selenium compounds and amino acid analysis, It is an acid protein with a pI value of 4.5-4.7 and stems from a cytosolic selenoprotein with an apparent native molecular mass of 240 kDa. It is present in various tissues, but is highly expressed in the epithelial cells of the prostate gland (Kalcklösch et al. 1995). The human 15 kDa selenoprotein was isolated from ^{75}Se-labeled human T cells and was shown to contain a selenocysteine residue encoded by TGA. Its pI value of 4.7 was calculated from its amino acid sequence. It has a similar native molecular mass of 200 to 240 kDa and is likewise highly expressed in the prostatic tissue (Gladyshev et al. 1998). We have recently been able to show that the 15 kDa selenoprotein detected in the rat and the human 15 kDa selenoprotein are the same protein in two different mammalian species (D. Röthlein et al. submitted) and therefore refer to both as 15 kDa selenoprotein (Sel15). Its function and its biological significance are not yet known but its further investigation is of special interest with regard to the decreased incidence of prostatic cancer with selenium supplementation (Clark et al. 1998). The recent finding that the gene for Sel15 is located on a chromosome often affected in cancer (Kumaraswamy et al. 2000) supports the hypothesis that this protein might play a role in the observed relationship between selenium and this disease.

5.3.4
18 kDa Selenoprotein

An 18 kDa selenium-containing protein was detected in various tissues of rats (Behne et al. 1988). It was identified as a selenocysteine-containing selenoprotein with an isoelectric point of about 4.6-4.8, which is mainly present in the mitochondrial membranes (Kyriakopoulos et al. 1996; Kyriakopoulos et al. 1999). Its biological function is still unknown. However, an interesting characteristic is the fact that in the hierarchy of selenium distribution it was found to be one of the most preferentially supplied proteins (Kyriakopoulos et al. 2002) which can be taken as an indication of its biological significance.

5.3.5
Further Selenoproteins Identified in Silico

Mammalian selenoprotein genes have been identified by scanning the nucleotide sequence databases for the selenocysteine insertion sequence elements necessary for decoding UGA as selenocysteine (Kryukov et al. 1999 Lescure et al. 1999). Nucleotide sequences corresponding to two novel selenoproteins, named SelR and SelT, have been found in one of the studies (Kryukov et al. 1999). The molecular masses of the human proteins, calculated from their cDNA sequences, are 12.6 kDa and 18.8 kDa for SelR and SelT respectively. The other study describes four novel selenoproteins SelN, SelX, SelZf1 and SelZf2, of which the latter two, TrxR2 homologs, have already been mentioned above (Lescure et al. 1999). Their calculated molecular masses are 58, 16 and 48 kDa for SelN, SelX and SelZ respectively. By means of Northern blot hybridization of human tissues, SelN mRNA was found to be ubiquitously expressed, but was enriched in pancreas, ovary, prostate and spleen, SelX mRNA was mainly present in liver and leukocytes and low in lung, placenta and brain, and SelZ mRNA was enriched in kidney, liver, testis and prostate and low in thymus. With the exception of the TrxR2 homologs no information on the functions of these selenoproteins is as yet available.

5.4
Selenium-Containing Proteins not yet Identified

Tracer experiments on animals labeled in vivo with ^{75}Se in combination with electrophoretic and chromatographic methods of protein separation (Hawkes et al. 1985; Danielson and Medina 1986; Calvin et al. 1987; Behne et al. 1988; Evenson and Sunde 1988). have been valuable tools in selenium rese-

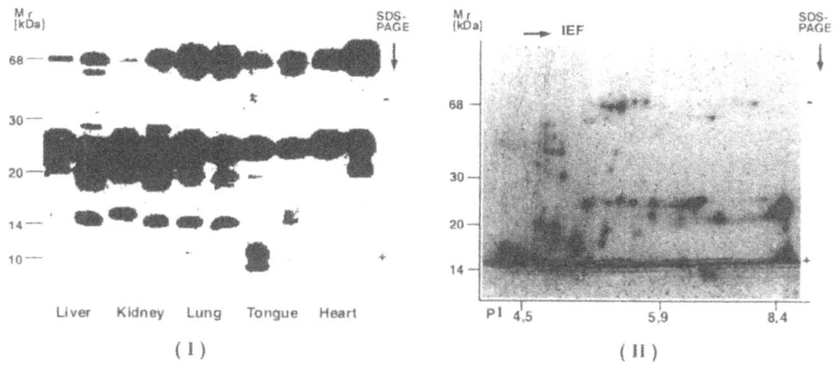

Fig. 1. Determination of selenium-containing proteins in rat tissues (Kyriakopoulos et al. 2001). (I) Autoradiogram of 75Se-labeled proteins or protein subunits in the homogenates of liver, kidney, lung tongue and heart after administration of 75Se-selenite to rats with a normal (left lane) or a low selenium status (right lane). (II) 75Se-labeled proteins in the kidney homogenate of a selenium-deficient rat after two-dimensional protein separation by SDS-PAGE / isoelectric focusing. In this way the labeled bands found after SDS-PAGE were resolved into more than 25 selenium-containing spots with molecular masses between 10 and 80 kDa and isoelectric points in the range between 3 and 9.

arch which, with most of the mammalian selenoproteins identified so far, gave a first indication of their existence. In studies on rats it has been shown that the incorporation of the labeled element into specific proteins was the more effective the lower the selenium status of the animals and the smaller the amount of the element administered (Behne et al. 1988). Thus selenium-depleted rats and a ^{75}Se-labeled tracer with a very high specific activity have been used. One advantage of this approach is that in this way maximum specific activity of the tracer in the proteins can be achieved. Furthermore it has been found that in selenium deficiency the metabolized element is excreted to a much lesser extent and instead is reutilized and redistributed among the body pools (Behne and Höfer-Bosse 1984). After an initiation phase the tracer activity therefore reflects the distribution of the native selenium. The third advantage is that due to the hierarchy in selenoprotein expression the incorporation of the element into the main selenoprotein, the cytosolic glutathione peroxidase, is greatly reduced, which results in the enhanced labeling of the other selenoproteins (Behne et al. 1988). It is thus possible to detect selenium-containing proteins present in the tissues at only very low concentrations. By combining this labeling procedure with high resolution protein separation methods such as SDS-PAGE and two-dimensional isoelectric focusing / SDS-PAGE, numerous selenium-containing proteins have been detected in rat tissues, as can be seen in the examples shown in Fig. 1. In the separation by SDS-PAGE 28 selenium-containing bands with molecular masses in the range from 116 to 8 kDa could be distinguished (Behne et al. 1996b). This range was extended by applying a modified tricine-SDS-PAGE, which allows the determination of smaller proteins. In this way four additional selenium-containing proteins with molecular masses between 7 kDa and 3 kDa have been detected (Kyriakopoulos et al. 2000). By means of two-dimensional SDS-PAGE / isoelectric focusing some of the selenium-containing bands could be further resolved into several spots with different isoelectric points as has been shown, for example, for rat kidney homogenate (Fig. 1b). Although a few of the ^{75}Se-labeled compounds may be precursors or metabolic products of the same selenoprotein, and selenium-binding proteins have also to be taken into account, the comparison with the selenoproteins so far known indicated that in addition to the selenocysteine-containing proteins detected so far there are further mammalian selenoproteins awaiting identification.

6 Conclusions and Outlook

Although with the advanced methods in molecular biology and protein biochemistry considerable progress has recently been made in selenium research, our knowledge of the selenium-containing proteins present in the organisms and of their biological significance is still far from complete and varies largely for the different forms of life. In yeast, for instance, no essential effects of selenium have been observed, and the genetic components needed for specific selenocysteine insertion into proteins have not been detected. However, a selenocysteine-containing protein has been described which exhibits glutathione peroxidase activity (Haas and Velten 1992) . Further studies are therefore needed to clarify the question of the existence of selenoproteins in yeast cells. The situation is very similar with regard to the role of selenium in the plant kingdom. Although plants constitute one of the major pathways, via which inorganic selenium enters the biosphere, they themselves do not seem to require the element for their life processes, and no solid evidence for the presence of specific selenoproteins in higher plants has been found. The detection of selenium-containing tRNAs (Wen et al. 1988; Hatfield et al. 1992) suggests, however, that plants might be able to synthesize specific selenoproteins, and this should encourage further research in this field.

The greatest progress has been made in research on prokaryotic and mammalian selenoproteins. 'In silico' analysis of the genetic elements for selenocysteine insertion has considerably increased our knowledge of novel selenoproteins in prokaryotic (Wilting et al. 1997) as well as in mammalian cells (Lescure et al. 1999; Kryukov et al. 1999). Using this approach several novel selenoproteins have also been detected in other eukaryotic organisms, such as drosophila (Martin-Romero et al. 2001) and zebrafish (Kryukov and Gladyshev 2000). Labeling experiments on rats showed the existence of further mammalian selenium-containing proteins (Behne et al. 1996b), and alternative splicing may also increase the number of selenoproteins which differ with regard to their biological tasks (Behne et al. 2001, Sun et al. 2001).

It is interesting to note that, with the exception of selenophosphate synthetase 2, selenium has different functions in eukaryotes and prokaryotes. So far the eukaryotic homologs of prokaryotic selenoproteins were found to lack selenocysteine, and vice versa. In this respect it has to be clarified whether selenium-containing proteins do exist in the eukaryotes in which, similar to some of the prokaryotic proteins, selenium is biologically active in forms other than selenocysteine.

Because of its nutritional and medical implications the research on the selenium-containing proteins present in the mammalian organisms is of particular interest. Here, in addition to the functional characterization of the novel selenoproteins, the elucidation of their biological roles in the different tissues is a great challenge. The significance of the element for mammals has been emphasized by the finding that the total disruption of selenoprotein synthesis by knocking out the selenocysteyl-tRNA gene was lethal in mice at the embryonic stage (Bösl et al. 1997). The fact that similar effects have not been observed in selenium deficiency is most probably due to the hierarchy in selenium distribution (Behne et al. 1988, Behne et al. 1999) and shows that the levels of preferentially supplied selenoproteins cannot be lowered sufficiently to enable their biological roles to be investigated by means of depletion experiments. Gene knockout studies will therefore be the best approach to obtain information on the functions and effects of a certain selenoprotein, independent of the regulation of the selenoprotein metabolism and interference from other selenium compounds.

Another important task in selenium research is the elucidation of the hierarchy in the distribution of the element among tissues and selenium-containing compounds in periods of insufficient selenium supply, but also when higher doses of the element are given. The resolution of this question will make it possible to settle the long-lasting discussion about optimum selenium intake and to obtain information on the compounds and mechanisms responsible for the protective effects of additional selenium supplementation.

References

Akasaka M, Mizoguchi J, Takahashi K (1990) A human cDNA sequence for a novel glutathione peroxidase-related protein. Nucleic Acids Res 18: 4619

Andreesen JR, Ljungdahl LG (1973) Formate dehydrogenase of *Clostridium thermoaceticum*, incorporation of selenium-75, and the effects of selenite, molybdate, and tungstate on the enzyme. J Bacteriol 116: 867-873

Andreesen JR, Wagner M, Sonntag D, Kohlstock M, Harms C, Gursinsky T, Jäger J, Parther T, Kabisch U, Gräntzdörffer A, Pich, A, Söhling B (1999) Various functions of selenols and thiols in anaerobic Gram-positive, amino acids-utilizing bacteria. BioFactors 10: 263-370

Arai M, Imai H, Sumi D, Imanaka T, Takano T, Chiba N, Nakagawa Y (1996) Import into mitochondria of phospholipid hydroperoxide glutathione peroxidase requires a leader sequence. Biochem Biophys Res Commun 227: 433-439

Arkowitz RA, Abeles RH (1989) Identification of acetyl phosphate as the product of clostridial glycine reductase: evidence for an acyl enzyme intermediate. Biochemistry 28: 4639-4644

Arkowith RA, Abeles RH (1990) Isolation and characterization of a covalent selenocysteine intermediate in the glycine reductase system. J Am Chem Soc 112: 870-872

Arteel GE, Mostert V, Oubrahim H, Briviba K, Abel J, Sies H (1998) Protection by selenoprotein P in human plasma against peroxynitrite-mediated oxidation and nitration. Biol Chem 379: 1201-1205

Arthur JR, Nicol F, Beckett GJ (1990) Hepatic iodothyronine 5′-deiodinase. The role of selenium. Biochem J 272: 537-540

Avissar N, Ornt DB, Yagil Y, Horowitz S, Watkins RH, Kerl EA, Takahashi K, Palmer IS, Cohen HJ (1994) Human kidney proximal tubules are the main source of plasma glutathione peroxidase. Am J Physiol 266: C367-C375

Axley MJ, Böck A, Stadtman TC (1991) Catalytic properties of an Escherichia coli formate dehydrogenase mutant in which sulfur replaces selenium. Proc Natl Acad Sci USA 88: 8450-8454

Bansal MP, Cook RG, Danielson KG, Medina D (1989) A 14 kilodalton selenium-binding protein in mouse liver is fatty acid-binding protein. J Biol Chem 264: 13780-13784

Bansal MP, Mukhopadhyay T, Scott J, Cook RG, Medina D (1990) DNA sequencing of a mouse liver protein that binds selenium: implications for selenium's mechanism of action in cancer prevention. Carcinogenesis 11: 2071-2073

Beck MA, Kolbeck PC, Shi Q, Rohr LH, Morris VC, Levander OA (1994) Increased virulence of a human enterovirus (coxsackievirus B3) in selenium-deficient mice. J Infect Dis 170: 351-357

Beck MA, Esworthy RS, Ho YS, Chu FF (1998) Glutathione peroxidase protects mice from viral-induced myocarditis. FASEB J 12: 1143-1149

Behne D, Höfer H, von Berswordt-Wallrabe R, Elger W (1982) Selenium in the testis of the rat: Studies on its regulation and its importance for the organism. J Nutr 112: 1682-1687

Behne D, Wolters W (1983) Distribution of selenium and glutathione peroxidase in the rat. J Nutr 113: 456-461

Behne D, Höfer-Bosse T (1984) Effects of a low selenium status on the distribution and retention of selenium in the rat. J Nutr 114: 1289-1296

Behne D, Duk M, Elger W (1986) Selenium content and glutathione peroxidase activity in the testis of the maturing rat. J Nutr 116: 1442-1447

Behne D, Hilmert H, Scheid S, Gessner H, Elger W (1988) Evidence for specific selenium target tissues and new biologically important selenoproteins. Biochim Biophys Acta 966: 12-21

Behne D, Kyriakopoulos A, Meinhold H, Köhrle J (1990a) Identification of type I iodothyronine 5'-deiodinase as a selenoenzyme. Biochem Biophys Res Commun 173: 1143-1149

Behne D, Scheid S, Kyriakopoulos A, Hilmert H (1990b) Subcellular distribution of selenoproteins in the liver of the rat. Biochem Biophys Acta 1033: 219-225

Behne D, Kyriakopoulos A, Scheid S, Gessner H. (1991) Effects of chemical form and dosage on the incorporation of selenium into tissue proteins in rats. J Nutr 121: 806-814

Behne D, Weiler H, Kyriakopoulos A (1996a) Effects of selenium deficiency on testicular morphology and function in rats. J Repr 106: 291-297

Behne D, Kyriakopoulos A, Weiss-Nowak C, Kalcklösch M, Westphal C, Gessner H (1996b) Newly found selenium-containing proteins in the tissues of the rat. Biol Trace Elem Res 55: 99-110

Behne D, Kyriakopoulos A, Kalcklösch M, Weiss-Nowak C, Pfeifer H, Gessner H, Hammel C (1997) Two new selenoproteins found in the prostatic glandular epithelium and in the spermatid nuclei. Biomed Environm Sci 10: 340-345

Behne D, Pfeifer H, Röthlein D, Kyriakopoulos A (2000) Cellular and subcellular distribution of selenium and selenium-containing proteins in the rat. In: Roussel AM, Favier AE, Anderson RA (eds) Trace Elements in Man and Animals 10, Kluwer Academic/Plenum Publishers New York, pp 29-34

Bermano G, Arthur JR, Hesketh JE (1996) Selective control of cytosolic glutathione peroxidase and phospholipid hydroperoxide glutathione peroxidase mRNA stability by selenium supply. FEBS Lett 387: 157-160

Berry MJ, Kieffer JD, Harney JW, Larsen PR (1991a) Selenocysteine confers the biochemical properties of the type I iodothyronine deiodinase. J Biol Chem 266: 14155-14158

Berry MJ, Banu L, Larsen PR (1991b) Type I iodothyronine deiodinase is a selenocysteine-containing enzyme. Nature 349: 438-440

Berry MJ, Banu L, Harney JW, Larsen PD (1993) Functional characterization of the eukaryotic SECIS elements which direct selenocysteine insertion at UGA codons. EMBO J 12: 3315-3322

Björnstedt M, Xue J, Huang W, Akesson B, Holmgren A (1994) The thioredoxin and glutaredoxin systems are efficient electron donors to human plasma glutathione peroxidase. J Biol Chem 269: 29382-29384

Böck A, Forchhammer K, Heider J, Leinfelder W, Sawers G, Veprek B, Zinoni F (1991a) Selenocysteine: the 21st amino acid. Mol Microbiol 5: 515-520

Böck A, Forchhammer K, Heider J, Baron C (1991b) Selenoprotein synthesis: an expansion of the genetic code. Trends Biochem Sci 16: 463-467

Bösl MR, Takaku K, Oshima M, Nishimura S, Taketo MM (1997) Early embryonic lethality caused by targeted disruption of the mouse selenocysteine tRNA gene (Trsp) Proc Natl Acad Sci USA 94: 5531-5534

Bottino NR, Banks CH, Irgolic KJ, Micks P, Wheeler AE, Zingaro RA (1984) Selenium-containing amino acids and proteins in marine algae. Phytochemistry 23: 2445-2452

Boyington JC, Gladyshev VN, Khangulov SV, Stadtman TC, Sun PD (1997) Crystal structure of formtc dehydrogenase H: catalysis involving Mo, molybdopterin, selenocysteine, and an Fe_4S_4 cluster. Science 275: 1305-1308

Broderick SJ, Deagen JT, Whanger PD (1987) Properties of the glutathione peroxidase isolated from human plasma. J Inorg Biochem 30: 299-308
Brown TA, Shrift A (1980) Identification of selenocysteine in the proteins of selenate-grown *Vigna radiata*. Plant Physiol 66: 758-761
Brown TA, Shrift A (1982) Selenium: toxicity and tolerance in higher plants. Biol Rev 57: 59-84
Broyer TC, Lee DC, Asher CJ (1966) Selenium nutrition of green plants. Effect of selenite supply on growth and selenium content of alfalfa and subterranean clover. Plant Physiol 41: 1425-1428
Broyer TC, Johnson CM, Huston RP (1972) Selenium and nutrition of *Astragalus*. 1. Effect of selenite or selenate supply on growth and selenium content. Plant and Soil 36: 635-649
Burk RF (1973) Effect if dietary selenium on ^{75}Se–binding to rat plasma proteins. Proc. Soc. Exp. Biol. Med. 143: 719-722
Burk RF, Hill KE, Awad JA, Morrow JD, Kato T, Cockell KA, Lyons PR (1995) Pathogenesis of diquat-induced liver necrosis in selenium-deficient rats. Assessment of the roles of lipid peroxidation by measurement of F2 isoprostanes. Hepatology 21: 561-569
Butler JA, Beilstein MA, Whanger PD (1989) Influence of dietary methionine on the metabolism of selenomethionine in rats. J Nutr 119: 1001-1009
Calvin HI, Cooper GW, Wallace E (1981) Evidence that selenium in rat sperm is associated with a cysteine-rich structural protein of the mitochondrial capsules. Gamete Res 4: 139-149
Calvin HI, Grosshans K, Musicant-Shikora SR, Turner SI (1987) A developmental study of rat sperm and testis selenoproteins. J Repr Fertil 81: 1-11
Casalone E, di Ilio C, Federiei G, Polsinelli M (1988) Glutathione and glutathione metabolizing enzymes in yeasts. Antonie van Leeuwenhoek 54: 367-375
Chambers I, Frampton J, Goldfarb P, Affara N, McBain W, Harrison P R (1986) The structure of the mouse glutathione peroxidase gene: the selenocysteine in the active site is encoded by the „termination codon“, TGA. EMBO J 5: 1221-1227
Cheng, WH, Ho Y-S, Valentine BA, Ross DA, Combs GF, Lei XG (1998) Cellular glutathione peroxidase is the mediator of body selenium to protect against paraquat lethality in transgenic mice. J Nutr 128: 1070-1076
Christensen MJ, Burgener KW (1992) Dietary selenium stabilizes glutathione peroxidase mRNA in rat liver. J Nutr 122: 1620-1626
Chu FF, Esworthy RS, Doroshow JH, Doan K, Liu XF (1992) Expression of plasma glutathione peroxidase in human liver in addition to kidney, heart, lung, and breast in humans and rodents. Blood 79: 3233-3238
Chu FF, Doroshow JH, Esworthy RS (1993) Expression, characterization, and tissue distribution of a new cellular selenium-dependent glutathione peroxidase, GSHPx-GI. J Biol Chem 268: 2571-2576
Chu FF, Esworthy RS, Ho YS, Swiderek K, Elliiot RW (1997) Expression and chromosomal mapping of mouse Gpx2 gene encoding the gastrointestinal form of glutathione peroxidase, GPX-GI. Biomed Environ Sci 10: 156-162
Clark LC, Dalkin B, Krongrad A, Combs GF Jr, Turnbull BW, Slate EH, Witherington R, Herlong JH, Janosko E, Carpenter D, Borosso C, Falk S, Rounder J (1998) Decreased incidence of prostate cancer with selenium supplementation: results of a double-blind cancer prevention trial. Br J Urol 81: 730-734
Combs GF Jr, Combs SB (1986a) The role of selenium in nutrition. Academic Press, Orlando Florida, pp 127-177
Combs GF Jr, Combs SB (1986b) The role of selenium in nutrition. Academic Press, Orlando Florida, pp 265-399

Cone JE, Martin del Rio R, Davis J N, Stadtman T C (1976) Chemical characterization of the selenoprotein component of clostridial glycine reductase: identification of selenocysteine as the organoselenium moiety. Proc Natl Acad Sci USA 73: 2659-2663

Copeland PR, Fletcher JE, Carlson BA, Hatfield DL, Driscoll DM (2000) A novel RNA binding protein, SPB2, is required for the translation of mammalian selenoprotein mRNAs. EMBO J 19: 306-314

Cox JC, Edwards ES, DeMoss JA (1981) Resolution of distinct selenium-containing formate dehydrogenases from *Escherichia coli*. J Bacteriol 145: 1317-1324

Croteau W, Whittemore SL, Schneider MJ, St Germain DL (1995) Cloning and expression of a cDNA for a mammalian type III iodothyronine deiodinase. J Biol Chem 270: 16569-16575

Croteau W, Davey JC, Galton VA, St Germain DL (1996) Cloning of the mammalian type II iodothyronine deiodinase. J Clin Invest 98: 405-417

Danielson KG, Medina D (1986) Distribution of selenoproteins in mouse mammary epithelial cells in vitro and in vivo. Cancer Res 46: 4582-4589

Davey JC, Becker KB, Schneider MJ, St. Germain DL, Galton VA (1995) Cloning of a cDNA for the type II iodothyronine deiodinase. J Biol Chem 270: 26786-26789

Deagen JT, Butler JA, Beilstein MA, Whanger PD (1987) Effects of dietary selenite, selenocystine and selenomethionine on selenocysteine lyase and glutathione peroxidase activities and on selenium levels in rat tissues. J Nutr 117: 91-98

de Haan JB, Bladier C, Griffiths P, Kelner M, O'Shea RD, Cheung NS, Bronson RT, Silvestro MJ,Wild S, Zheng SS, Beart PM, Hertog PJ, Kola I (1998) Mice with a homozygous null mutation for the most abundant glutathione peroxidase, Gpx1, show increased susceptibility to the oxidative stress-inducing agents paraquat and hydrogen peroxide. J Biol Chem 273: 22528-22536

Dilworth GL (1982) Properties of the selenium-containing moiety of nicotinic acid hydroxylase from Clostridium barkeri. Arch Biochem Biophys 219: 30-38

Dobbek H, Gremer L, Meyer O, Huber R (1999) Crystal structure and mechanism of CO dehydrogenase, a molybdo iron-sulfur flavoprotein containing S-selanylcysteine. Proc Natl Acad Sci USA 96: 8884-8889

Doucette GJ, Price NM, Harrison PJ (1987) Effects of selenium deficiency on the morphology and ultrastructure of the coastal marine diatom *Thalassiosira preudonana (Bacillariaphyceae)*. J Phycol 23: 9-17

Drotar A, Phelps P, Fall R (1985) Evidence for glutathione peroxidase activities in cultured plant cells. Plant Sci 42: 35-40

Dürre P, Andreesen JR (1983) Purine and glycine metabolism by purinolytic clostridia. J Bacteriol 154: 192-199

Easwari K and Lalitha K (1995) Subcellular distribution of selenium during uptake and its influence on mitochondrial oxidations in germinating *Vigna radiata L.* Biol Trace Elem Research 48: 141-160

Eidsness MK, Scott RA, Prickril BC, DerVartanian DV, Legall J, Moura I, Moura JJG, Peck HD Jr (1989) Evidence for selenocysteine coordination to the active site nickel in the (NiFeSe) hydrogenases from Desulfovibrio baculatus. Proc Natl Acad Sci USA 86: 147-151

Enoch HG, Lester RL (1975) The purification and properties of formate dehydrogenase and nitrate reductase from *Escherichia coli*. J Biol Chem 250: 6693-6705

Esworthy RS, Swiderek KM, Ho YS, Chu FF (1998) Selenium-dependent glutathione peroxidase-GI is a major glutathione peroxidase activity in the mucosal epithelium of rodent intestine. Biochim Biophys Acta 1381: 213-226

Evenson JK, Sunde RA (1988) Selenium incorporation into selenoproteins in the Se-adequate and Se-deficient rat. Proc Soc Exp Biol Med 187: 169-180
Fauque G, Peck HD Jr, Moura JJ, Huynh BH, Berlier Y, DerVartanian DV, Teixeira M, Przybyla AE, Lespinat PA, Moura I, LeGall J (1988) The three classes of hydrogenases from sulfate-reducing bacteria of the genus *Desulfibria*. FEMS Microbiol Rev 54: 299-344
Flohé L, Günzler WA, Schock HH (1973) Glutathione peroxidase: a selenoenzyme. FEBS Letters 32: 132-134
Forchhammer K, Böck A (1991) Selenocysteine synthase from Escherichia coli. Analysis of the reaction sequence. J Biol Chem 266: 6324-2628
Forstrom JW, Zakowski JJ, Tappel AL (1978) Identification of the catalytic site of rat liver glutathione peroxidase as selenocysteine Biochemistry 17: 2639-2644
Franke K W, Potter W R (1935) A new toxicant occurring naturally in certain samples of plant foodstuffs. IX. Toxic effects of orally ingested selenium. J Nutr 10: 213-222
Galiazzo F, Schiesser A, Rotilio G (1987) Glutathione peroxidase in yeast. Presence of the enzyme and induction by oxidative conditions. Biochem Biophys Res Commun 147: 1200-1205
Garcia E, Vernede X, Hatchikian EC, Volbeda A, Frey M, Fontecilla-Camps JC (1999) The crystal structure of a reduced [NiFeSe] hydrogenase provides an image of the activated catalytic center. Structure Fold Des 7: 557-566
Gasdaska PY, Gasdaska JR, Cochran S, Powis G (1995) Cloning and sequencing of a human thioredoxin reductase. FEBS Lett 373: 5-9
Gasdaska JR, Harney JW, Gasdaska PY, Powis G, Berry MJ (1999a) Regulation of human thioredoxin reductase expression and activity by 3'-untranslated region selenocysteine insertion sequence and mRNA instability elements. J Biol Chem 274: 25379-25385
Gasdaska PY, Berggren MM, Berry ML, Powis G (1999b) Cloning, sequencing and functional expression of a novel human thioredoxin reductase. FEBS Lett 442: 105-111
Gladyshev VN, Khangulov SV, Stadtman TC (1996a) Properties of the selenium- and molyb-denum-containing nicotinic acid hydroxylase from Clostridium barkeri. Biochemistry 35: 212-223
Gladyshev VN, Jeang KT, Stadtman TC (1996b) Selenocysteine, identified as the penultimate C-terminal residue in human T-cell thioredoxin reductase, corresponds to TGA in the human placental gene. Proc Natl Acad Sci USA 93: 6146-6151
Gladyshev VN, Jeang KT, Wootton JC, Hatfield DL (1998) A new human selenium-containing protein. Purification, characterization, and cDNA sequence. J Biol Chem 273: 8910-8915
Gladyshev VN, Krause M, Xu XM, Korotkov KV, Kryukov GV, Sun Q A Lee BJ, Wootton JC, Hatfield DL (1999) Selenocysteine-containing thioredoxin reductase in C.elegans. Biochem Biophys Res Commun 259: 244-249
Grahame DA (1988) A summary of new findings in research on hydrogenase. BioFactors 1: 279-283
Gromer S, Wissing J, Behne D, Ashman K, Schirmer H, Flohé L, Becker K (1998) A hypothesis on the catalytic mechanism of the selenoenzyme thioredoxin reductase. Biochem J 332: 591-592
Gu QP, Beilstein MA, Barofsky E, Ream W, Whanger PD (1999) Purification, characterization and glutathione binding to selenoprotein W from monkey muscle. Arch Biochem Biophys 361: 25-33

Günzler WA, Steffens GJ, Grossmann A, Kim SMA, Ötting F, Wendel A, Flohé L (1984) The amino-acid sequence of bovine glutathione peroxidase. Hoppe-Seyler's Z Physiol Chem 365: 195-212

Guimaraes MJ, Peterson D, Vicari A, Cocks BG, Copeland NG, Gilbert DJ, Jenkins, NA, Ferrick DA, Kastelein RA, Bazan JF, Zlotnik A (1996) Identification of a novel selD homolog from eukaryotes, bacteria and archaea: is there an autoregulatory mechanism in selenocysteine metabolism? Proc Natl Acad Sci USA 93: 15086-15091

Haas HJ, Velten M (1992) Selenoproteins in mitochondria and cytosol of *Saccharomyces uvarum* after growth in sodium selenite- supplemented media. J Trace Elem Electrolytes Health Dis 6: 71-74

Halboth S, Klein A (1992) *Methanococcus* voltae harbors four gene clusters potentially encoding two (NiFe) and two (NiFeSe) hydrogenases, each of the cofactor F_{420}-reducing or F_{42}-non-reducing types. Mol Gen Genet 233: 217-224

Hammel C, Kyriakopoulos A, Behne D, Gawlik D, Brätter P (1996) Protein-bound selenium in the seeds of coco de mono (*Lecythis ollaria*). J Trace Elem Med Biol 10: 96-102

Harrison PJ, Yu PW, Thompson PA, Price NM, Phillips DJ (1988) Survey of selenium requirements in marine phytoplankton. Mar Ecol Prog Ser 47: 89-96

Hatfield DL, Lee BJ, Price NM, Stadtman TC (1991) Selenocysteyl-tRNA occurs in the diatom *Thalassiosira* and in the ciliate *Tetrahymena*. Mol Microbiol 5: 1183-1186

Hatfield D, Choi IS, Mischke S, Owens LD (1992) Selenocysteyl-tRNAs recognize UGA *in Beta vulgaris*, a higher plant, and in *Gliocladium virens*, a filamentous fungus. Biochem Biophys Res Commun 184: 254-259

Hawkes WC, Wilhelmsen EC, Tappel A (1985) Abundance and tissue distribution of selenocysteine-containing proteins in the rat. J Nutr 113: 456-461

Heider J, Forchhammer K, Sawers G, Böck A (1991) Interspecies compatibility of selenoprotein biosynthesis in *Enterobacteriaceae*. Arch Microbiol 155: 221-228

Heider J, Baron C, Böck A (1992) Coding from a distance: dissection of the mRNA determinants required for the incorporation of selenocysteine into protein. EMBO J 11: 3759-3766

Herrman JL (1977) The properties of a rat serum protein labelled by the injection of sodium selenite. Biochim Biophys Acta 500: 61-70

Hill KE, Lloyd RS, Yang JG, Read R, Burk RF (1991) The cDNA for rat selenoprotein P contains 10 TGA codons in the open reading frame. J Biol Chem 266: 10050-10053

Ho Y-S, Magnenat J-L, Bronson RT, Cao J, Gargano M, Sugawara M, Funk CD (1997) Mice deficient in cellular glutathione peroxidase develop normally and show no increase sensitivity to hyperoxia. J Biol Chem 272: 16644-16651

Holmgren A, Björnstedt M (1995) Thioredoxin and thioredoxin reductase. Methods Enzymol. 252 B: 199-208

Huber RE, Criddle RS (1967) Comparison of chemical properties of selenocysteine and selenocystine with their sulfur analogs. Arch Biochem Biophys 122: 164-173

Imhoff D, Andreesen JR (1979) Nicotinic acid hydroxylase from Clostridium barkeri: selenium-dependent formation of active enzyme. FEMS Microbiol Lett 5: 155-158

Jones JB, Dilworth GL, Stadtman TC (1979) Occurrence of selenocysteine in the selenium-dependent formate dehydrogenase of *Methanococcus vannielii*. Arch Biochem Biophys 195: 255-260

Jones JB, Stadtman TC (1981) Selenium-dependent and selenium-independent formate dehydrogenases of *Methanococcus vannielii*. J Biol Chem 256: 656-663

Kabisch UC, Gräntzdörffer A, Schierhorn A, Rücknagel KP, Andreesen JR, Pich A (1999) Identification of D-proline reductase from *Clostridium sticklandii* as a selenoenzyme and indications for a catalytically active pyruvoyl group derived from a cysteine residue by cleavage of a proprotein. J Biol Chem 274: 8445-8454

Kalcklösch M, Kyriakopoulos A, Hammel C, Behne D (1995) A new selenoprotein found in the glandular epithelial cells of the rat prostate. Biochem Biophys Res Commun 217: 162-70

Kaplan MM (1986) Regulatory influences on iodothyronine deiodination in animal tissues. In: Hennemann G (ed) Thyroid Hormone Metabolism, Marcel Dekker New York, pp 231-253

Keshan Disease Research Group (1979) Observations on effect of sodium selenite in prevention of Keshan disease. Chinese Med J 92: 471-476

Kopf A, Brätter P, Gramm HJ (1990) Spurenelementkonzentrationen bei langzeitparenteraler Ernährung unter besonderer Berücksichtigung des Selenmetabolismus. Anaesthesist 39:633-634

Korhola M, Vainio A, Edelmann K (1986) Selenium yeast. Ann Clin Res 18: 65-68

Kreimer S, Andreesen JR (1995) Glycine reductase of Clostridium litorale. Cloning, sequencing and molecular analysis of the grdAB operon that contains two in-frame TGA codons for selenium incorporation. Eur J Biochem 234: 192-199

Kryukov GV, Kryukov VM, Gladyshev VN (1999) New mammalian selenocysteine-containing proteins identified with an algorithm that searches for selenocysteine insertion sequence elements. J Biol Chem 274: 33888-33897

Kryukov GV, Gladyshev VN (2000) Selenium metabolism in zebrafish: multiplicity of selenoprotein genes and expression of a protein containing 17 selenocysteine residues. Genes Cells 5: 1049-1060

Kumaraswamy E, Malykh A, Korotkov KV, Kozyavkin S, Hu Y, Kwon SY, Moustafa ME, Carlson BA, Berry MJ, Lee BJ, Hatfield DL, Diamond AM, Gladyshev VN (2000) Structure-expression relationships of the 15-kDa selenoprotein gene. Possible role of the protein in cancer etiology. J Biol Chem 275: 35540-35547

Kyriakopoulos A, Kalcklösch M, Weiß-Nowak C, Behne D (1993) Studies on 16 kDa selenium-containing proteins enriched by means of preparative electrophoresis. Electrophoresis 14: 108-111

Kyriakopoulos A, Hammel C, Gessner H, Behne D (1996) Characterization of an 18 kDa-selenium-containing protein in several tissues of the rat. Am Biotech Lab14: 22

Kyriakopoulos A, Pfeifer H, Hammel C, Röthlein D, Geßner H, Behne D (1997) Investigation of nuclear selenium-containing proteins in the kidney of the rat. In: Anke M et al. (eds) Mengen- und Spurenelemente, 17. Arbeitstagung, Verlag Schubert H, Leipzig, pp 438-446

Kyriakopoulos A, Röthlein D, Pfeifer H, Geßner H, Behne D (1998) Selenium-containing compounds in baker´s yeast (Saccharomyces cerevisiae). In: Anke M, Meißner D (eds) Mengen- und Spurenelemente, 18. Arbeitstagung, Verlag Schubert H, Leipzig, pp 740-745

Kyriakopoulos A, Röthlein D, Pfeifer H, Geßner H, Behne D (1999) Distribution of an 18 kDa-selenoprotein in the mitochondrial membranes in several tissues of the rat. In: Anke M et al. (eds) Mengen- und Spurenelemente, 19. Arbeitstagung, Verlag Schubert H, Leipzig, pp 292-301

Kyriakopoulos A, Röthlein D, Pfeifer H, Bertelsmann H, Kappler S, Behne D (2000) Detection of small selenium-containing proteins in tissues of the rat. J Trace Elements Med Biol 14: 170-183

Kyriakopoulos A, Bertelsmann H, Kappler S, Behne D (2001) Investigation of membrane-bound selenium-containing proteins in cell compartments of several tissues of the rat. In: Anke M, Müller R, Schäfer U (eds) Mineralstoffe. Mengen-, Spuren- und Ultraspurenelemente in der Prävention. Wissenschaftliche Verlagsgesellschaft, Stuttgart, pp 374-380

Kyriakopoulos A, Bertelsmann H, Graebert A, Hoppe B, Kühbacher M, Behne D (2002) Distribution of an 18 kDa selenoprotein in several tissues of the rat. J Trace Elements Med Biol 16: 57-67

Lacourciere GM (1999) Biosynthesis of selenophosphate. BioFactors 10: 237-244

Lee BJ, Worland PJ, Davis JN, Stadtman TC, Hatfield DL (1989) Identification of a selenocysteyl tRNASer in mammalian cells that recognizes the nonsense codon UGA. J Biol Chem 264: 9724-9727

Lee SR, Kim JR, Kwon KS, Yoon HW, Levine RL, Ginsburg A, Rhee SG (1999) Molecular cloning and characterization of a mitochondrial selenocysteine-containing thioredoxin reductase from rat liver. J Biol Chem 274: 4722-4734

Lee SR, Bar-Noy S, Kwon J, Levine RL, Stadtman TC, Rhee SG (2000) Mammalian thioredoxin reductase: Oxidation of the C-terminal cysteine/selenocysteine active site forms a thioselenide, and replacement of selenium with sulfur markedly reduces catalytic activity. Proc Natl Acad Sci USA 97: 2521-2526

Leinfelder W, Zehelein E, Mandrand-Berthelot MA, Böck A (1988) Gene for a novel tRNA species that accepts L-serine and cotranslationally inserts selenocysteine. Nature 331: 723-725

Leinfelder W, Forchhammer K, Veprek B, Zehelein E, Böck A (1990) In vitro synthesis of selenocysteyl-$tRNA^{UCA}$ from seryl-$tRNA^{UCA}$: involvement and characterization of the selD gene product. Proc Natl Acad Sci USA 87: 543-547

Leonard DM, Stachelek SJ, Safran M, Farwell AP, Kowalik TF, Leonard JL (2000) Cloning, expression, and functional characterization of the substrate binding subunit of rat type II iodothyronine 5'-deiodinase. J Biol Chem 275: 25194-25201

Leonhardt U, Andreesen J R (1977) Some properties of formate dehydrogenase, accumulation and incorporation of 185W-tungsten into proteins of *Clostridium formicoacetum*. Arch Microbiol 115: 277-284

Lescure A, Gautheret D, Carbon P, Krol A (1999) Novel selenoproteins identified in silico and in vivo by using a conserved RNA structural motif. J Biol Chem 274: 38147-38154

Li CZ, Huang JR, Li CX (1986) Sodium selenite as a preventive measure for Kaschin-Beck disease as evaluated in X-ray studies. In:Combs GF Jr, Spallholz JE, Levander OA, Oldfield JE (eds) Selenium in Biology and Medicine, AVI Press New York, pp 934-937

Low SC, Harney JW, Berry MJ (1995) Cloning and functional characterization of human selenophosphate synthetase, an essential component of selenoprotein synthesis. J Biol Chem 270: 21659-21664

Low SC, Berry MJ (1996) Knowing when not to stop: Selenocysteine incorporation in eukaryotes. Trends Biochem Sci 21: 203-208

Maddipati KR, Marnett IJ (1987) Characterization of the major hydroperoxide reducing activity of human plasma. J Biol Chem262: 17398-17403

Maiorino M, Wissing JB, Brigelius-Flohé R, Calabrese F, Roveri A, Steinert P, Ursini F, Flohé L (1998) Testosterone mediates expression of the selenoprotein PHGPx by induction of spermatogenesis and not by direct transcriptional gene activation. FASEB J 12: 1359-1370

Martin-Romero FJ, Kryukov GV, Lobanov AV, Carlson BA, Lee BJ, Gladyshev VN, Hatfield DL (2001) Selenium metabolism in Drosophila: selenoproteins, selenoprotein mRNA expression, fertility, and mortality (2001) J Biol Chem 276: 29798-29804

Matsui M, Oshima M, Oshima H, Takaku K, Maruyama T, Jodoi J, Taketo MM (1996) Early embryonic lethality caused by targeted disruption of the mouse thioredoxin gene. Dev Biol 178: 179-185

McCoy KEM and Weswig PH (1969) Some selenium response in the rat not related to vitamin E. J Nutr 98: 383-389

Meyer M, Granderath K, Andreesen JR (1995) Purification and characterization of protein PB of betaine reductase and its relationship to the corresponding proteins glycine reductase and sarcosine reductase from Eubacterium adidaminophilum. Eur J Biochem 234: 184-191

Millar KR (1972) Distribution of ^{75}Se in liver, kidney, and blood proteins of rats after intravenous injection of sodium selenite. NZJ Agr Res 15: 547-564

Mills GC (1957) Hemoglobin catabolism. I. Glutathione peroxidase, an erythrocyte enzyme which protects hemoglobin from oxidative breakdown. J Biol Chem 229: 189-197

Miranda-Vizuete A, Damdimopoulos AE, Pedrajas JR, Gustafsson JA, Spyrou G (1999) Human mitochondrial thioredoxin reductase. Eur J Biochem 261: 405-412

Mortimer RH, Galligan JP, Cannell GR, Addison RS, Roberts MS (1996) Maternal to fetal thyroxine transmission in the human term placenta is limited by inner ring deiodination. J Clin Endocrinol Metab 81: 2247-2249

Motsenbocker MA, Tappel AL (1982) A selenocysteine-containing selenium-transport protein in rat plasma. Biochim Biophys Acta 719: 147-153

Mustacich D, Powis G (2000) Thioredoxin reductase. Biochem J 346: 1-8

Muth E, Morschel E, Klein A (1987) Purification and characterization of an 8-hydroxy-5-deazaflavin-reducing hydrogenase from the archaebacterium *Methanococcus voltae*. Eur J Biochem 169: 571-577

Neuhierl B, Böck A (1996) On the mechanism of selenium tolerance in selenium-accumulating plants. Purification and characterization of a specific selenocysteine methyltransferase from cultured cells of *Astragalus bisculatus*. Eur J Biochem 239: 235-238

Patterson E L, Milstrey R, Stokstad E L R (1957) Effect of selenium in preventing exudative diathesis in chicks. Proc Soc Exp Biol Med 95: 617-620

Peck HD, Gest H (1957) Formic dehydrogenase and the hydrogenlyase enzyme complex in coli-aerogenes bacteria. J Bacteriol 73: 706-721

Peterson PJ, Butler GW (1962) The uptake and assimilation of selenite by higher plants. Aust J Biol Sci 15: 126-146

Pinsent J (1954) The need for selenite and molybdate in the formation of formic dehydrogenase by members of the *Coli-aerogenes* group of bacteria. Biochem J 57: 10-16

Pfeifer H, Conrad M, Roethlein D, Kyriakopoulos A, Brielmeier M, Bornkamm GW, Behne D (2001) Identification of a specific sperm nuclei selenoenzyme necessary for protamine thiol cross-linking during sperm maturation. FASEB J 15: 1236-1238

Price NM, Thompson PA, Harrison PJ (1987) Selenium: an essential element for growth of the coastal marine diatom *Thalassiosira pseudonana (Bacillariophyceae)*. J Phycol 23: 1-9

Price NM, Harrison PJ (1988) Specific selenium-containing macromolecules in the marine diatom *Thalassiosira pseudonana*. Plant Physiol 86: 192-199

Pushpa-Rekha TR, Burdsall AL, Oleksa LM, Chisolm GM, Driscoll DM (1995) Rat phospholipid-hydroperoxide glutathione peroxidase. cDNA cloning and identification of multiple transcription and translation start sites. J Biol Chem 270: 26993-26999

Read R, Bellow T, Yang J-G, Hill KE, Palmer IS, Burk RF (1990) Selenium and amino acid composition of selenoprotein P , the major selenoprotein in rat serum. J Biol Chem 265: 17899-17905

Rotruck JT, Pope AL, Ganther HE, Swanson AB, Hafeman DG, Hoekstra WG (1973) Selenium: biochemical role as a component of glutathione peroxidase. Science 179: 588-590

Roveri A, Casaco A, Maiorino M, Dalan P, Calligaro A, Ursini F (1992) Phopholipid hydroperoxide glutathione peroxidase of rat testis: gonadotropin dependency and immunocytochemical identification. J Biol Chem 267: 6142-6146

Roveri A, Maiorino M, Ursini F (1994) Enzymatic and immunological measurements of soluble and membrane-bound PHGPx. Methods Enzymol 233: 202-212

Saijoh K, Saito N, Lee MJ, Fujii M, Kobayashi T, Sumino K (1995) Molecular cloning of cDNA encoding a bovine selenoprotein P-like protein containing 12 selenocysteines and a (His-Pro) rich domain insertion, and its regional expression. Mol Brain Res 30: 301-311

Salbe AD, Levander OA (1990) Effect of various dietary factors on the deposition of selenium in the hair and nails of rats. J Nutr 120: 200-206

Salvatore D, Bartha T, Harney JW, Larsen PR (1996) Molecular biological and biochemical characterization of the human type 2 selenodeiodinase. Endocrinology 137: 3308-3315

Sani BP, Woodard JL, Pierson MC, Allen RD (1988) Specific binding proteins for selenium in rat tissues. Carcinogenesis 9: 277-84

Sathe SK, Mason AC, Weaver CM (1992a) Some properties of a selenium-incorporating sulfur-rich protein in soybeans *(Glycine max L.)*. J Agric Food Chem 40: 2077-2083

Sathe SK, Mason AC, Rodibaugh R, Weaver CM (1992b) Chemical form of selenium in soybean *(Glycine max L.)* lectin. J Agric Food Chem 40: 2084-2091

Sawers G, Heider J, Zehelein E, Böck A (1991) Expression and operon structure of the sel genes of *Escherichia coli* and identification of a third selenium-containing formate dehydrogenase isoenzyme. J Bacteriol 173: 4983-4993

Schräder T, Rienhofer A, Andreesen JR (1999) Selenium-containing xanthine dehydrogenase from Eubacterium barkeri. Eur J Biochem 264: 862-871

Schwarz K, Foltz C M (1957) Selenium as an integral part of factor 3 against dietary necrotic liver degeneration. J Am Chem Soc 79: 3292-3293

Self WT, Stadtman TC (2000) Selenium-dependent metabolism of purines: A selenium-dependent purine hydroxylase and xanthine dehydrogenase were purified from Clostridium purinolyticum and characterized. Proc Natl Acad Sci USA 97: 7208-7213

Shigeoka S, Takeda, T, Hanaoka T (1991) Characterization and immunological properties of selenium-containing glutathione peroxidase induced by selenite *in Chlamydomonas reinhardtii*. Biochem J 275: 623-625

Soehling B, Parther T, Ruecknagel KP, Wagner M, Andreesen JR (2001) A selenocysteine-containing peroxiredoxin from the strictly anaerobic organism Eubacterium acidaminophilum. Biol Chem 282 (in press)

Sorgenfrei O, Duin EC, Klein A, Albracht SP (1997) Changes in the electronic structure around Ni in oxidized and reduced selenium-containing hydrogenases from Methanococcus voltae. Eur J Biochem 247: 681-687

Sorgenfrei O, Linder D, Karas M, Klein A(1993) A novel very small subunit of a selenium containing [NiFe] hydrogenase of Methanococcus voltae is posttranslationally processed by cleavage at a defined position. Eur J Biochem 213: 1355-1358

Sun QA, Wu Y, Zappacosta F, Jeang KT, Lee BJ, Hatfield DL, Gladyshev VN (1999) Redox regulation of cell signaling by selenocysteine in mammalian thioredoxin reductases. J Biol Chem 274: 24522-24530

Sun QA, Zappacosta F, Factor VM, Wirth PJ, Hatfield DL, Gladyshev VN (2001) Heterogeneity within animal thioredoxin reductases: evidence for alternative first exon splicing. J Biol Chem 276: 3106-3114
Sun Y, Ha PC, Butler JA, Ou BR, Yeh JY, Whanger P (1998) Effect of dietary selenium on selenoprotein W and glutathione peroxidase in 28 tissues of the rat. J Nutr Biochem 9: 23-27.
Sun Y, Gu QP, Whanger PD (2001) Selenoprotein W in overexpressed and underexpressed rat glial cells in culture. J Inorg Biochem 84: 151-156
Sunde RA, Evenson JK (1987) Serine incorporation into the selenocysteine moiety of glutathione peroxidase. J Biol Chem 262: 933-937
Sunde RA, Thompson RM, Palm MD, Weiss SL, Thompson KM, Evenson JK (1997) Selenium regulation of selenium-dependent glutathione peroxidases in animals and transfected CHO cells. Biomed Environ Sci 10: 346-355
Takahashi K, Avissar N, Whitin J, Cohen H (1987) Purification and characterization of human plasma glutathione peroxidase: a selenoglycoprotein distinct from the known cellular enzyme. Arch Biochem Biophys 256: 677-686
Tamura T, Stadtman TC (1996) A new selenoprotein from human lung adenocarcinoma cells: Purification, properties, and thioredoxin reductase activity. Proc Natl Acad Sci USA 93: 1006-1011
Tanaka H, Stadtman TC (1979) Selenium-dependent clostridial glycine reductase. Purification and characterization of the two membrane-associated protein components. J Biol Chem 254: 447-452
Teixeira M, Fauque G, Moura I, Lespinat PA, Berlier Y, Prickril B, Peck HD Jr, Xavier AV, LeGall J, Moura JJG (1987) Nickel-(iron-sulfur)-selenium containing hydrogenases from *Desulfovibrio baculatus* (DSM 1743) Eur J Biochem 167: 47-58
Thomas JP, Maiorino M, Ursini F, Girotti AW (1990) Protective action of phospholipid hydroperoxide glutathione peroxidase against membrane-damaging lipid peroxidation. In situ reduction of phospholipid and cholesterol hydroperoxides. J Biol Chem 265: 454-461
Trelease SF, Trelease HM (1939) Physiological differentiation in Astragalus with reference to selenium. Amer J Bot 26: 530-535
Tujebajeva RM, Copeland PR, Xu XM, Carlson BA, Harney JW, Driscoll DM, Hatfield DL, Berry MJ (2000) Decoding apparatus for eukaryotic selenocysteine insertion EMBO Reports 1: 158-163
Turner DC, Stadtman TC (1973) Purification of protein components of the clostridial glycine reductase system and characterisation of protein A as a selenoprotein. Arch Biochem Biophys 154: 366-381
Ursini F, Maiorino M, Valente M, Ferri L, Gregolin C (1982) Purification from pig liver of a protein which protects liposomes and biomembranes from peroxidative degradation and exhibits glutathione peroxidase activity on phopatidylcoline hydroperoxides. Biochim Biophys Acta 710: 197-211
Ursini F, Maiorino M, Gregolin C (1985) The selenoenzyme phospholipid hydroperoxide glutathione peroxidase. Biochim Biophys Acta 839: 62-70
Ursini F, Heim S, Kiess M, Maiorino M, Roveri A, Wissing J, Flohé L (1999) Dual function of the selenoprotein PHGPx during sperm maturation. Science 285: 1393-1396
Vendeland SC, Beilstein MA, Chen CL, Jensen ON, Barofsky E, Whanger PD (1993) Purification and properties of selenoprotein-W from rat muscle. J Biol Chem 268: 17103-17107
Vendeland SC, Beilstein MA, Yeh JY, Ream W, Whanger PD (1995) Rat skeletal muscle selenoprotein W: cDNA clone and mRNA modulation by dietary selenium. Proc Natl. Acad Sci USA 92: 8749-8753

Veres Z, Tsai L, Scholz TD, Politino M, Balaban, RS, Stadtman TC (1992) Synthesis of 5-methylaminomethyl-2-selenouridine in tRNAs: ^{31}P NMR studies show the labile selenium donor synthesized by the selD gene product contains selenium bonded to phosphorus. Proc Natl Acad Sci USA 89: 2975-2979

Veres Z, Kim IY, Scholz TD, Stadtman TC (1994) Selenophosphate synthetase. Enzyme properties and catalytic reaction. J Biol Chem 269: 10597-10603

Volk DM, Cutliff SA (1986) Selenium deficiency and cardiomyopathy in a patient with cystic fibrosis. J Ky Med Assoc 84: 222-224

Voordouw G, Menon NK, LeGall J, Choi ES, Peck HD Jr, Przybyla AE (1989) Analysis and comparison for (NiFe) and (NiFeSe) hydrogenases from *Desulfovibrio gigas* and *Desulfovibrio baculatus.* J Bacteriol 171: 2894-2899

Vorholt JA, Vaupel M, Thauer RK (1997) A selenium-dependent and a selenium-independent formylmethanofuran dehydrogenase and their transcriptional regulation in the hyperthermophilic *Methanopyrus kandleri.* Mol Microbiol 23: 1033-1042

Wagner R, Andreesen JR (1977) Differentiation between *Clostridium acidiurici* and *Clostridium cylindrosporum* on the basis of specific metal requirements for formate dehydrogenase formation. Arch Microbiol 114: 219-224

Wagner R, Andreesen JR (1979) Selenium requirement for active xanthine dehydrogenase from Clostridium acidiurici and Clostridium cylindrosporum. Arch Microbiol 121: 255-260

Wagner M, Sonntag D, Grimm R, Pick A, Eckerskorn C, Söhling B, Andreesen JR (1999) Substrate-specific selenoprotein B of glycine reductase from Eubacterium acidaminophilum. Biochemical and molecular analysis. Eur J Biochem 260; 38-49

Waschulewski IH, Sunde RA (1988) Effect of dietary methionine on tissue selenium and glutathione peroxidase (EC 1.11.1.9) activity in rats given selenomethionine. Br J Nutr 60: 57-68

Watabe S, Makino Y, Ogawa K, Hiroi T, Yamamoto Y, Takahashi SY (1999) Mitochondrial thioredoxin reductase in bovine adrenal cortex. Its purification, nucleotide/amino acid sequence, and identification of selenocysteine. Eur J Biochem 264: 74-84

Wendel A, Kerner B, Graupe A (1978) The selenium moiety of glutathione peroxidase. Hoppe-Seyler's Z Physiol Chem 359: 1035-1036

Whanger PD, Butler JA (1988) Effects of various dietary levels of selenium as selenite or selenomethionine on tissue selenium levels and glutathione peroxidase activity in rats. J Nutr 118: 846-852

Wingler K, Böcher M, Flohé L, Kollmus H, Brigelius-Flohé R (1999) mRNA stability and SECIS efficiency rank gastrointestinal glutathione peroxidase high in the hierarchy of selenoproteins. Eur J Biochem 259: 149-157

Wilting R, Schorling S, Persson BC, Böck A (1997) Selenoprotein synthesis in archaea: Identification of an mRNA element of Methanococcus jannaschii probably directing selenocysteine insertion. J Mol Biol 266: 637-641

Wu SH, Oldfield JE, Whanger PD, Weswig PH (1969) Effect of selenium on reproduction. Proc West Sec Am Soc Anim Sci 20: 85-89

Yamada K, Nakagawa CW, Mutoh N (1999) Schizosaccharomyces pombe homologue of glutathione peroxidase, which does not contain selenocysteine, is induced by several stresses and works as an antioxidant. Yeast 15: 1125-1132

Yamamoto I, Saiki T, Liu SM Ljungdahl LG (1983) Purification and properties of NADP-dependent formate dehydrogenase from *thermoaceticum Clostridium,* a tungsten-selenium-iron protein. J Biolog Chem 258: 1826-1832

Yamazaki S (1982) A selenium-containing hydrogenase from *Methanocosus vannielii.* J Biol Chem 257: 7926-7929

Yamazaki S (1987) Properties of selenium-containing hydrogenase *from Methanococcus vannielii*. In: Combs GF Jr, Spallholz JE, Levander OA, Oldfield JE (eds) Selenium in Biology and Medicine, AVI Press New York, pp 230-235

Yokota A, Shigeoka S, Onishi T, Kitaoka S (1988) Selenium as inducer of glutathione peroxidase in low-CO2-grown *Chlamydomonas reinhardtii*. Plant Physiol 86: 649-651

Yoshimura S, Watanabe K, Suemizu H, Onozawa T, Mizoguchi J, Tsuda K, Katta H, Moriuchi T (1991) Tissue specific expression of the plasma glutathione peroxidase gene in rat kidney. J Biochem 109: 918-23

Zinoni F, Birkmann A, Stadtman TC, Böck A (1986) Nucleotide sequence and expression of the selenocysteine-containing polypeptide of formate dehydrogenase (formate-hydrogen-lyase-linked) from *Escherichia coli*. Proc Natl Acad Sci USA 83: 4650-4654

Constitutively Active and G-Protein coupled inward rectifier K$^+$ channels: Kir2.0 and Kir3.0

Peter R. Stanfield[1], Shigehiro Nakajima[2] and Yasuko Nakajima[3]

[1]Molecular Physiology Group, Department of Biological Sciences, University of Warwick, Coventry CV4 7AL, UK. [2]Department of Pharmacology, and [3]Department of Anatomy and Cell Biology, University of Illinois at Chicago, College of Medicine, Chicago, Illinois 60612, USA

Contents

1 Introduction 50
2 Classification and structure 52
2.1.1 Sub-families of inward rectifier K$^+$channels 52
2.1.2 Primary structure of Kir channels 52
2.1.3 Accessory proteins 61
3 Consitutively active strong inward rectifiers belong to the Kir2.0 subfamily 62
3.1 Inward rectifiers are activated by extracellular K$^+$ 62
3.2 Inward rectifier K$^+$ channels are multi-ion pores 63
3.2.1 Voltage dependent blockage by extracellular foreign cations 63
3.2.2 Measurements of flux ratio; anomalous mole fraction effects 64
3.3 Gating of inward rectifiers 65
3.3.1 Role of intracellular Mg^{2+} 65
3.3.2 Mg^{2+} and substate behaviour 65
3.3.3. Gating by intracellular polyamines 66
3.3.4 Interactions of Mg^{2+} and polyamines 73
3.3.5 Measured affinities for blockage by polyamines 73
3.3.6 Residues that form the receptor for gating polyamines 75
3.3.7 Kir channels may show intrinsic gating 80
3.3.8 Comparison with N-type inactivation in Kv 83
3.4 Constitutive activity requires phosphatidylinositol 4,5-bisphosphate 84
3.5 Dependence on $[K^+]_0$ 86
3.6 Selectivity and blockage of inward rectifier K$^+$ channels 92
3.6.1 Role of the selectivity filter 92
3.6.2 Blockage by foreign cations is influenced by several sites through the pore 96
3.7 Modulation of channels of the Kir2.0 subfamily 97

Reviews of Physiology, Biochemistry, and Pharmacology, Vol. 145

4 G-protein coupled inward rectifier K^+ channels belong to the Kir3.0 subfamily 103
4.1 Atrial K_{ACh} 104
4.2 Gβγ activates Kir3.0 105
4.3 Kir3.0: Heterotetramers 106
4.4 Distribution of Kir2.0 and Kir3.0 in the brain 108
4.5 Properties of activated Kir3.0 channels 109
4.5.1 Fast bursts and slow bursts 109
4.5.2 Molecular correlates of long and short openings 111
4.6 Mechanisms of Rectification in Kir3.0 Channels 112
4.7 The Gate of Kir3.0 Channels 112
4.8 Mechanisms of Gβγ-Kir3.0 Interaction 113
4.8.1 Gβγ Molecule Binds Kir3.0 113
4.8.2 Which Kir3.0 Domains Functionally Interact with Gβγ? 114
4.8.3 Which Gβγ sites interact with Kir3.0? 115
4.8.4 Gγ and Kir3.0 channels 117
4.8.5 Regulators of G Protein Signalling (RGS) 117
4.9 Specificity of Signal Transduction in Kir3.0 Activation 118
4.9.1 Specificity: Interaction of G protein and Kir3.0 119
4.9.2 Specificity: Association of Receptor and G Protein 119
4.9.3 Specificity: Physical Proximity 120
5. Modulators of Kir3.0 124
5.1 PIP_2 and ATP 124
5.2 Intracellular Na^+ 125
5.3 Synaptic Potentials and Channel Modulation 126
5.3.1 Slow Synaptic Potentials 126
5.3.2 Slow Inhibitory Synaptic Potentials in Neurones 127
5.3.3 Gβγ Activation of Neuronal Kir3.0 127
5.3.4 Slow Excitation in Nucleus Basalis Neurons 128
5.3.5 Slow Excitation in Locus Coeruleus and Other Neurons 129
5.3.6 Inhibition of Kir3.0 Channels in Atrial Myocytes and Hetorologous Expression Systems 130
5.3.7 Comparison of Kir3.0 and KirNB 133
5.3.8 Other Examples of Kir Modulation 133
6 Physiological function of strong inward rectifiers 134
6.1 Resting potential and the activation and de-activation of Kirs by K^+_o 137
6.1.1 Resting potential of skeletal muscle 137
6.1.2 Expression of Kir channels – resting potential and development 137
6.1.3 K^+ acts as a vasodilator by activating Kir channels and enhancing K^+ efflux 138

6.2 Kir channels contribute to action potential shape and duration: Kir2.0 of heart muscle........................ 141
6.3 K^+ homeostasis - K^+ activation of Kir channels and K^+ influx .. 143
6.3.1 Requirements for entry of K^+ through Kir............. 144
6.3.2 K^+ homeostasis in the nervous system................ 145
6.3.3 Kir and actions potential in skeletal muscle............ 146
6.4 Concluding comment
Abbreviations 148
References 149

1 Introduction

Inward rectifier K^+ channels (Kir channels) make up a family of ion channels found in the membranes of many cell types. The molecular structure of these channels is simple, channels being formed as tetramers of subunits that have just two membrane spanning domains. The channels behave as rectifiers in the terminology that an electrical engineer would use. Thus, when the direction of K^+ flux is outward, from cytosol to extracellular fluid, Kir channels tend to close; but when K^+ flux is inward, Kir channels open. This mechanism enables cells not only to conserve K^+ but also to facilitate K^+ entry into the cell.

The several potassium channels families now identified may be grouped into two major classes: those gated by voltage (Kv) and those showing inward rectification (Kir). Cole and Curtis (1941) first showed that depolarization applied to squid axon results in a larger membrane conductance than does hyperpolarization. This non-linear behaviour, the first example of Kv, was called "delayed rectification" (Hodgkin *et al.*, 1949). In contrast, Katz (1949) observed that when skeletal muscle fibres are immersed in high K^+ solution, the membrane conductance is larger over hyperpolarizing potentials. This unexpected behaviour was first called *anomalous rectification* (translated from the French, in which the first description was published, *propriétés détectrices anormales*') but is now more commonly called *inward rectification* (Adrian *et al.*, 1970).

From the beginning, the phenomenon of inward rectification was paradoxical. Measurements of K^+ fluxes using a radioisotope revealed that even in skeletal muscle fibres, there is an increase in K^+ permeability during depolarization (Hodgkin and Horowicz, 1959a). This finding was puzzling because it appeared to contradict the results of Katz (1949). Hodgkin and Horowicz (1959a) speculated "that depolarization leads to an increase in potassium permeability in both muscle and nerve, but that in the former tissue the rise lasts for only a few ms, instead of being maintained as long as the membrane is depolarized, as it is in nerve". This speculation proved to be true. Indeed the delayed rectification in muscle is transient, and is completely inactivated by a few seconds of depolarization (Nakajima *et al.*, 1962).

Hodgkin and Horowicz (1959b) further noticed that the activation of the inward rectifier is unusual and fundamentally different from that of the delayed rectifier. Activation of Kir does not depend on membrane potential *per se* and the channels are not voltage-gated. Rather, activation depends on the difference between the membrane potential and the equilibrium potential for K^+ $(V - E_K)^*$. This was a striking insight. More than 30 years later, the structure of Kir channels was shown to be fundamentally different from that of voltage-gated K^+ channels (Papazian *et al.*, 1987, Tempel *et al.*, 1987; Kubo *et al.*, 1993a). Kir channels lack the voltage-gating machinery of Kv, and are gated by intracellular substances such as Mg^{2+} and polyamines (Matsuda *et al.*, 1987; Lopatin *et al.*, 1994).

During the 1960s and 1970s, the biophysical properties of Kir channels were elucidated. Activation was shown to depend on $(V - E_K)$ if, but only if, $[K^+]_o$, not $[K^+]_i$, were changed (Hagiwara & Yoshii, 1979; Leech & Stanfield, 1981). Inward rectification was attributed to block of the channel by particles from the cytosol (Armstrong, 1969; Hille and Schwarz, 1978; Standen and Stanfield, 1978b). Biophysical analyses were performed on blockage of Kir channels by ions such as Ba^{2+}, Cs^+, and Na^+, mainly using skeletal muscle fibres and oocytes (Standen and Stanfield, 1978a; Hagiwara and Jaffe, 1979). Initially, only a "classical", constitutively active inward rectifier was known. Subsequent research has revealed other types of Kir channels. The ACh-activated Kir channel (K_{ACh} or G protein-coupled Kir) was found in atrial cells of the heart (Sakmann *et al.*, 1983); the ATP-sensitive Kir channel (K_{ATP}) in heart muscle cells (Noma, 1983); and epithelial, ATP-regulated Kir in kidney (Wang and Giebisch, 1991). These discoveries kindled interest in Kir channels as a biologically important class of ion channels.

Then in 1993, the cloning of Kir molecules began, with the sequencing of three types of Kir channels - rat outer medullary K^+ channel (ROMK1; Kir1.1; Ho *et al.*, 1993), the classical inward rectifier (IRK1; Kir2.1; Kubo *et al.*, 1993a), and the G-protein coupled inward rectifier K^+ channel (GIRK1; Kir3.1; Dascal *et al.*, 1993; Kubo *et al.*, 1993b). These events strongly stimulated the interest of many researchers in Kir channels, and the number of papers on the subject has risen sharply.

In 1998 Doyle *et al.* determined the crystal structure of a K^+ channel (KcsA K^+ channel) in bacteria *Streptomyces lividans*. This towering work finally gives us a means to understand functional results based on the quaternary structure of ion channels.

This review deals with channels belonging to the Kir2.0 and the Kir3.0 subfamilies. Other Kir channels are included only when general properties or comparisons are discussed. We emphasize the physiological roles of the channels; the last chapter deals with functional significance of the inwardly rectifying property of the channels. We will discuss how inward rectification helps the cell conserve K^+; in addition, we will deal with the mechanisms by which inward rectification facilitates the entry of K^+ into the cells, helping the cell to establish the normal ionic gradients. In this way we attempt to complement the several related reviews published recently, while seeking to avoid redundancy (Clapham, 1994; Clapham and Neer, 1997; Dascal, 1997; Doupnik *et al.*, 1995a; Isomoto *et al.*, 1997; Jan and Jan, 1994, 1997; Mark and Herlitze, 2000; Nichols and Lopatin, 1997; Wickman and Clapham, 1995; Yamada *et al.*, 1998).

* Footnote: V is the membrane potential, and E_K is the K^+ equilibrium potential:

$$E_K = \frac{RT}{F}\ln\frac{[K^+]_o}{[K^+]_i}$$

in which $[K^+]_o$ and $[K^+]_i$ are, respectively, the extracellular and intracellular K^+ concentrations, and R, T and F have their usual meanings.

2
Classification and Structure

In this section we summarise the diversity of Kir channels that has been uncovered using molecular techniques and give a general description of channel structure.

2.1.1
Sub-Families of Inward Rectifier K^+Channels

There are seven sub-families of Kir channels (Table 1), many of which have several members. The sub-families are denoted Kir1.0 to Kir7.0 in the nomenclature introduced by Chandy & Gutman (1993). The classification is based on homology or identity in terms of sequence, with channels ascribed to a single sub-family having a high degree of identity. The principal origin of diversity lies in the multiplicity of genes for Kir channels, but diversity also arises partly through alternative splicing (e.g. Kir1.1a-f; Kir3.2a-c). The degree of identity in channels of the Kir3.0 family is relatively low. However, as we describe later, these channels function only as heterotetramers in heart (Krapivinsky *et al.*, 1995a) and brain (Kofuji *et al.*, 1995; Velimirovic *et al.*, 1996; Liao *et al.*, 1996). Some authorities (*e.g.* Shuck *et al.*, 1994b) have placed channels of the Kir4.0 subfamily with Kir1.0; we have not followed this usage. Table 1 also gives common names of the ion channels (ROMK, IRK, GIRK, etc), some having more than one such name as a result of their having been sequenced in different laboratories or from different species or tissues. We shall use common names such as GIRK only where the authors of the primary papers favour them, or where their use may help avoid confusion.
The relationship between certain channels is indicated in the dendrogram of Fig. 1. The family has been ascribed the name *KCNJ* by the Human Genome Organisation. The names of the genes encoding the various channels are also given in Table 1, as are the chromosomal positions of these genes in humans.

2.1.2
Primary structure of Kir channels

Members of the Kir family were first sequenced by Ho *et al.* (1993), Kubo *et al.* (1993a, b) and Dascal *et al.*, (1993). Kyte-Doolittle (1982) hydrophobicity plots, such as those given for Kir2.1 and Kir3.1 in Fig. 2A, B, indicate only two membrane-spanning domains, M1 and M2. These domains are thought to be α-helical in structure. This prediction is supported by: the X-ray crystallographic study of the bacterial channel KcsA (Doyle *et al.*, 1998); FTIR spectroscopy of Kir (Brazier, Ramesh, Haris, Lee & Srai, 1998); scanning cysteine accessibility mutagenesis of M2 (Lu *et al.*, 1999a); and other scanning mutagenesis methods (Collins *et al.*, 1997; Minor *et al.*, 1999). Of these domains,

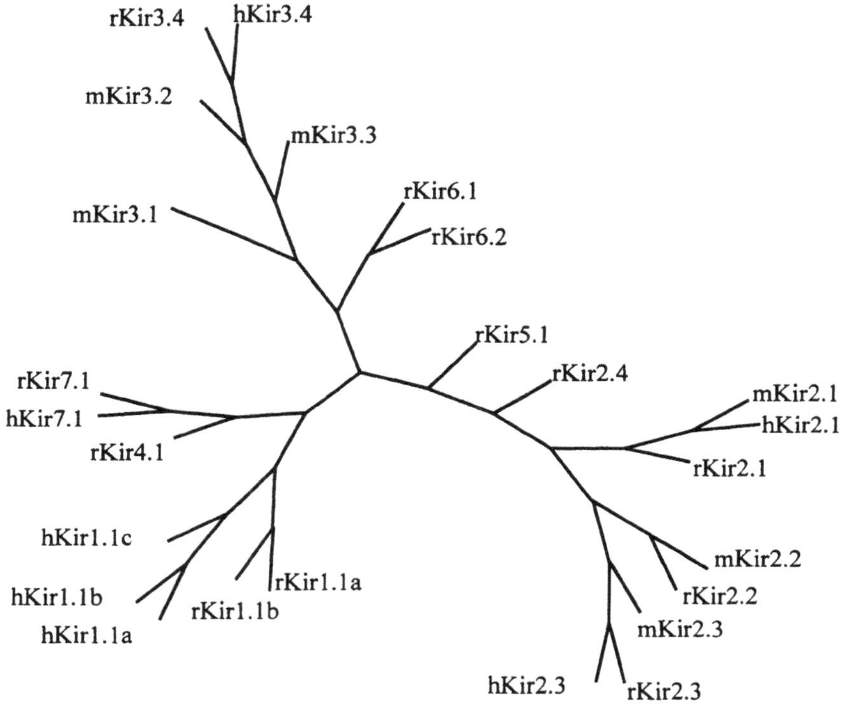

Fig. 1. Relationship among inward rectifer potassium channels and their subfamilies. Dendrogram illustrating relationships between certain members of the Kir channel gene family. h, human; m, mouse; r, rat. Accession numbers of protein sequences in the NCBI database are as follows: hKirl. 1a, NP_000211; rKirl. 1a,NP_058719; hKirl. 1b, XP_038165; rKirl. 1b, AAB30553; hKirl. 1c, XP_038164; hKir2.1, NP_000882; mKir2.1, NP_032451;rKir2.1, NP_058992; mKir2.2, NP_034733;rKir2.2, P52188; hKir2.3, NP_004972; mKir2.3, P52189; rKir2.3, P52190; rKir2.4, CAA05839; mKir3.1, NP_032452; mKir3.2, NP_034736; mKir3.3, NP_032455; hKir3.4,; rKir3.4, NP_058993; rKir4.1, CAA58568; rKir5.1, CAA58564; eKir6.1, NP_058795; rKir6.2, JC4689; hKir7.1, XP_044784; rKir7.1, 070617. (Courtesy of Dr MJ Sutcliffe, Department of Chemistry, University of Leicester.)

only M2 is likely to line the inner part of the pore (Doyle *et al.*, 1998; Minor *et al.*, 1999). And this inner part of the pore has a relatively wide diameter (>1.2nm at its widest) from the studies of channel blockage associated with the action of sulfhydryl reagents of the methanthiosulphonate class on substituted Cys residues within M2 (Lu *et al.*, 1999a). The inner pore is then described by Lu *et al.* (1999a) as being wider than that described in the bacterial K^+ channel, KcsA (Doyle *et al*, 1998). However, this difference may reflect measurements on channels that are open in the case of Kir2.1 and closed in

Table 1. Classification of inward rectifier K^+ channel (Kir) family

Subfamilies	Subtypes	Gene-name	*Cytogenic position (human)	Variants	References
Kir 1.0 (ATP-regulated K^+ channel)	Kir1.1a (ROMK1, K_{AB-1})	*KCNJ1*	11q24		Ho et al., 1993
				Kir1.1b-f (ROMK 2-6)	Shuck et al.,1994;Yano et al., 1994; Zhou et al., 1994; Boim et al., 1995, Kondo et al., 1996
Kir 2.0 (Resting inward rectifier K^+channel, Classical inward rectifier K^+channel)	Kir2.1 (IRK1)	*KCNJ2*	17q23.1-q24.2		Kubo et al., 1993a; Ishii et al., 1994; Raab-Graham et al., 1994
	Kir2.2 (IRK2)	*KCNJ12*	17p11.1		Koyama et al., 1994;Takahashi et al., 1994; Tang & Yang, 1994; Namba et al., 1996
				Kir2.2v	Namba et al., 1996
	Kir2.3 (IRK3, BIR 11)	*KCNJ4*	22q13.1		Bond et al., 1994; Lesage et al., 1994; Makhina et al., 1994; Morishige et al., 1994; Périer et al., 1994; Tang et al., 1994
	Kir2.4	*KCNJ14*	19q13		Töpert et al., 1998
Kir 3.0 (G protein-coupled inward rectifier K^+channel)	Kir 3.1 (GIRK1, KGA)	*KCNJ3*	2q24.1		Kubo et al., 1993b; Dascal et al., 1993; Kobayashi et al., 1995
				Kir3.1	Nelson et al., 1997; Zhu et al., 2001
	Kir 3.2 a (GIRK2, GIRK2A)	*KCNJ6*	21q22.13-q22.2		Lesage et al., 1994
				Kir3.2 variants	Bond et al., 1995; Ferrer et al., 1995; Stoffel et al., 1995; Tsaur et al., 1995; Isomoto et al., 1996; Wei et al., 1998; Inanobe et al., 1999

Table 1. Classification of inward rectifier K^+ channel (Kir) family

Subfamilies	Subtypes	Gene-name	*Cytogenic position (human)	Variants	References
	Kir 3.3 (GIRK3)	*KCNJ9*	1q21-q23		Lesage et al., 1994; Dissmann et al., 1996; Jelacic et al., 1999
	Kir 3.4 (GIRK4, CIR parenthesis	*KCNJ5*	11q24		Krapivinsky et al., 1995a; Lesage et al., 1995
	Kir 3.5 (GIRK5, XIR)				Hedin et al., 1996
Kir 4.0	Kir 4.1 (BIR10, $K_{AB-2,}$ Kir 1.2)	*KCNJ10*	1q22-q23		Bond et al., 1994; Takumi et al., 1995; Shuck et al., 1997
	Kir 4.2 (Kir 1.3)	*KCNJ15*	21		Gosset et al., 1997, Shuck et al., 1997; Pearson et al., 1999
Kir 5.0	Kir 5.1(BIR9)	*KCNJ16*	17q23.1-q24.1		Bond et al., 1994
Kir 6.0 (ATP-sensitive K^+ channel)	Kir.6.1 (u K_{ATP-1})	*KCNJ8*	12p11.23		Inagaki et al., 1995b
	Kir 6.2 (BIR)	*KCNJ11*	11p15.1		Inagaki et al., 1995a
Kir 7.0	Kir7.1	*KCNJ13*	2q37		Döring et al., 1998; Krapivinsky et al., 1998b; Partiseti et al., 1998

*Information from NCBI UniGene database

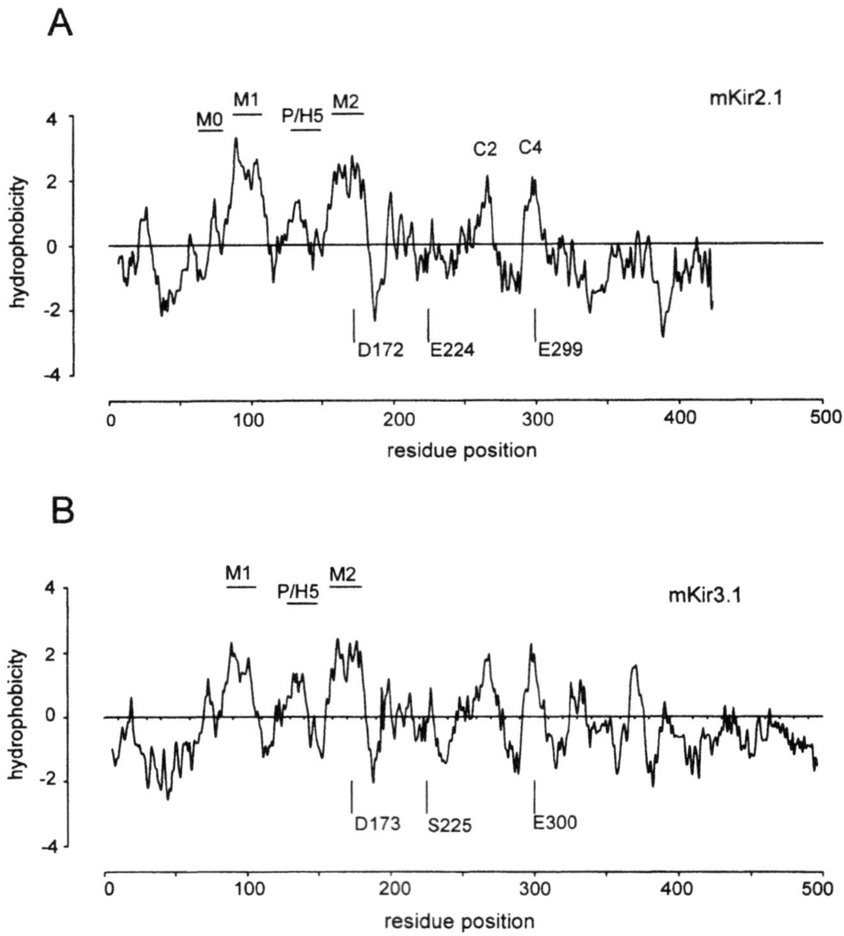
A
mKir2.1
M0
M1
P/H5
M2
C2
C4
D172
E224
E299
hydrophobicity
residue position
B
mKir3.1
M1
P/H5
M2
D173
S225
E300
hydrophobicity
residue position

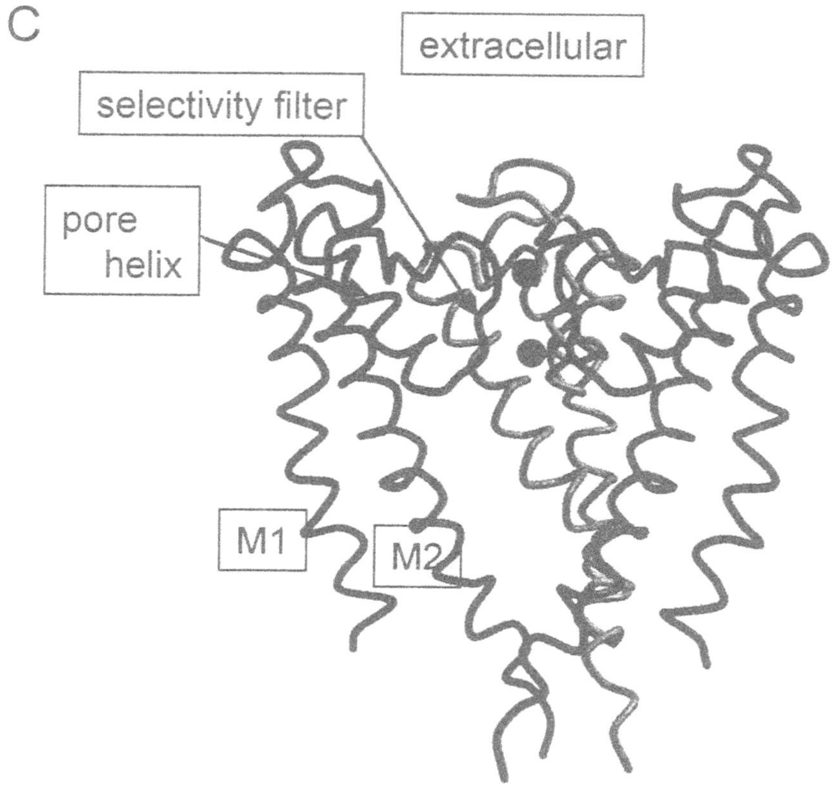

Fig. 2. Structure of inward rectifier potassium channels. A. Kyte-Doolitle hydropathy plot of murine Kir2.1 (Kyte & Doolitle, 1982) constructed using a window size of 11 residues. Regions M1 and M2 are believed to be membrane spanning; the region P/H5 froms the pore helix and selectivity filter of the ion channel. M0 has some similarity to S4 of Kv channels (Ho *et al.*, 1993). Hydrophobic regions indicated as C3 and C4 occur in all Kir channels and may be membrane associated. Residues D172, E224, and E299 are involved in gating of Kir2.1 by intracellular polyamines, as described in section 3 of the text. **B.** Kyte Doolitle hydropathy plot of murine Kir3.1, constructed using a window size of 11 residues. Regions M1, P/H5, and M2 ar indicated, as are the positions of the residues equivalent to those in Kir2.1 that are known to be involved in channel gating by poylamines. **C.** Structure of Kir2.1, based upon that of the bacterial channel KcsA (Doyle *et al.*, 1998), illustrating the peptide backbone of three of the four subunits that make up the ion channel. The subunit immediately in front of the pore has been removed for clarity. The structure of the N- and C-Termini is not illustrated. (Courtesy of Dr MJ Sutcliffe, Department of Chemistry, University of Leicester.)

the case of KcsA (see below & Fig. 3). It is in this inner, water filled part of the pore, lying about halfway across the membrane that a single hydrated K^+ ion resides, briefly awaiting its turn, during efflux, to enter the K^+ selectivity filter (*e.g.* Doyle *et al.*, 1998; Morais-Cabral *et al.*, 2001). Kir channels lack any equivalent of the domains S1 – S4 that confer voltage gating of Kv channels. However, as discussed in detail in the next sections, channels do nevertheless undergo conformation changes in response to appropriate stimuli.

A region of the N-terminal close to M1 has some homology to S4, but is not membrane spanning; this region was termed M0 by Ho *et al.* (1993). M0 has not been widely studied, though a Lys residue within it (K80) confers proton gating on ROMK1 over the physiological range of intracellular pH (Fakler *et al.*, 1996b). This Lys residue has its pK_a set by close proximity to arginine residues in the N- (R41) and C-terminus (R311) giving some clue as to folding of the these intracellular domains (Schulte *et al.*, 1999). Proton gating of Kir4.1-Kir5.1 heteromers appears to depend on a similar structure (Yang *et al.*, 2000). M0 is also important in the response of Kir2.3 to changes of intracellular pH (Qu *et al.*, 1999). Throughout the channel family, the region contains two conserved Asp residues; these are D71 and D78 in Kir2.1. Mutation of one of these residues to Val (D71V) is one cause of Andersen's syndrome in humans (Plaster *et al.*, 2001), a syndrome that includes abnormalities both of development and of excitability, particularly of heart (see section 6). This mutant acts as a dominant negative, preventing channel function in heteromers with wild type Kir2.1. Neither of the mutants D71C or D78C of Kir2.1 was active as a channel without concatenation with wild type Kir2.1 subunits (Lu *et al.*, 1999b).

As already indicated, studies of proton gating of ROMK1 suggest that the N- and C-terminus fold to come close to each other at the inner mouth of the pore (Schulte *et al.*, 2001). Yang *et al.* (1995a) and Taglialatela *et al.* (1995) also have evidence that part of the C-terminus close to, but beyond M2 contributes to the inner pore mouth since mutation of E224 in Kir2.1 alters unitary conductance and open channel current noise. Recent studies suggest that parts of the N- and C-termini form a long, wide intracellular vestibule. The evidence comes from replacement of several residues by Cys and from showing that such replacement permits at least partial pore blockage by the sulfhydryl reagents MTSEA (methanthiosulphonate ethylammonium) and MTSET (methanthiosulphonate ethyltrimethylammonium; Lu *et al.*, 1999b). The regions studied by Lu *et al.* (1999b) were from residues 54-86 in the N-terminus (*i.e.* including M0), where 15 mutants gave blockage by MTSEA or MTSET, and 213-234 in the C-terminus, where 11 mutants gave blockage. By building tandem tetramers with a varying number (0 to 4) of Cys substitutions at position E224, and by then probing with sulfhydryl reagents of differing sizes, Lu *et al.* (1999b) were able to gauge the volume of the inner vestibule. They show that it may be occupied by up to four molecules of the largest sulfhydryl reagent used (monobromtrimethylammoniobimane bromide); each of these molecules is 1.2 × 1.0 × 0.6nm. This inner vestibule may be akin to the 'hanging gondola' structure that forms the cytoplasmic vestibule of K^+ channels such as KcsA and Kv (Gulbis *et al.*, 2000; Kobertz *et al.*, 2000; Soklova *et al.*, 2001) and also of nicotinic acetylcholine receptors (Miyazawa *et al.* 1999).

Two hydrophobic regions are found in an equivalent position in the C terminus of all Kirs. These regions may be membrane associated or may be hidden from the aqueous environment by folding. They have been termed C2 and C4 by Taglialatela *et al.* (1995; Fig. 2A) in their work on the function of the C-terminus.

The H5 or P-region forms the selectivity filter of Kir channels, as in Kv and KcsA, and like them generally contains the TxxTxGYG motif that is the signature sequence of K^+ channels (Heginbotham *et al.*, 1994). Channels of the Kir6.0 subfamily, which form K_{ATP}, have GFG in place of GYG, as in K^+ channels of the *eag* family (*e.g.* Warmke & Ganetsky, 1994). With the exception of Kir7.1, the H5 region of Kir channels has an arginine residue just beyond the signature sequence (Table 4). As discussed in the next chapter, this residue appears fundamental in conferring the property of dependence on (V-E_K) first described by Hodgkin & Horowicz (1959b) or of activation by K^+_o (Hagiwara & Yoshii, 1979; Leech & Stanfield, 1981). It is this property that characterises the family, with the exception of Kir7.1.

The arrangement of M1 and M2 has been demonstrated in KcsA by X-ray crystallography. The helices form a structure likened to an inverted tepee by Doyle *et al.* (1998); M2 lines the pore; M1 makes contact with M2 and with lipid (Fig. 2C, Fig. 3). This arrangement appears to be that of the closed form of a K^+ channel. The arrangement of M1 and M2 has been described as different in Kir2.1 by Minor *et al.* (1999; Fig. 3). A study by Collins *et al* (1997) had used scanning Trp and Asp substitution to determine those residues that had aqueous (Asp) or lipid (Trp) contacts. This study found a surprisingly large tolerance of either substitution. Work by Minor *et al* (1999) tested the ability of mutant forms of Kir2.1 to rescue yeasts that otherwise lacked a K^+ transporter when grown in low [K^+] media. Large numbers of substitutions were made and yeast growth was used as an indicator of functional, open channels. The nature of substitutions that permitted K^+ transfer was used to predict whether the side chains of residues were exposed to water, protein or lipid. These studies indicated two surfaces for protein-protein interaction for both M1 and M2, with only one surface for M1 exposed to lipid. Thus M1 contacts two adjacent M2 domains in the K^+ channel multimer and *vice versa*. Similar studies on Kir3.0 channels (Yi *et al.*, 2001) suggest that the conformation described by Minor *et al.* (1999) may represent the open form of a K^+ channel, since Kir2.1 is constitutively active. In channels that open and close, movements of M2 are involved so that residues that are buried in the shut state become pore lining when open. For example replacement in Kir3.2 of a valine residue (V188, see Table 2) toward the C-terminal end of M2 by

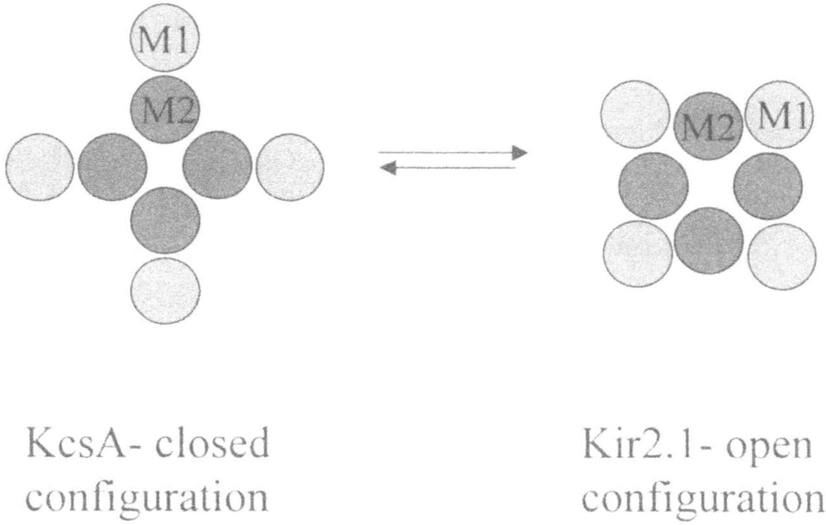

Fig. 3. Gating movements of Kir channels. The arrangement of M1 and M2 is illustrated for KcsA (at left; Doyle *et al.*, 1998) and Kir2.1 (at right; Minor *et al.*, 1999). Gating in Kir3.0 is believed to involve rotation of M2 so the shut state at right is converted into the open state at left (after Yi et al., 2001).

hydrophobic residues kept the channels shut, indicating that in this state the residue is buried (Yi *et al.*, 2001). Replacement by small, polar, or charged residues increased the probability that channels were open, suggesting that, in the open state, the residue now had its side chain in the pore. Similar conclusions have been reached for Kir3.0 by Sadja *et al.* (2001) and for Kir6.0 by Drain *et al.* (1998), Tucker *et al.* (1998) and Loussouarn *et al.* (2000). Thus, certain forms of channel gating in Kir involve conformation changes similar to those believed to occur in Kv (Liu *et al.*, 1997) or KcsA (Perozo *et al.*, 1999), with opening of the inner pore through movement of M2 (Fig. 3).

As already indicated, pore forming subunits come together as tetramers to form ion channels. This structure was initially proposed by analogy with Kv channels (Kubo *et al.*, 1993a), which are known to be tetrameric (*e.g.* Li *et al.*, 1994). Later, once it was established that intracellular polyamines such as spermine gate Kir channels and that certain residues act as receptors for these gating cations (*e.g.* Yang *et al.*, 1995a; see next section), polyamine affinity was used to establish Kir channels as tetramers. These experiments used mixtures of different wild type subunits with high and low affinity (Glowatski *et al.*, 1995) or of wild type and mutant subunits linked in tandem (Yang *et al*, 1995b). Other work using concatameric channels formed by linking subunits also confirms tetrameric structure (*e.g.* Yang *et al.*, 1995c; Dart *et al.*, 1998a; Lu *et al.*, 1999a, b). Kir channels lack the N-terminal tetramerisation domain found in Kv (Li *et al.*, 1992) and certainly M0 does not function in this way. Rather the proximal C-terminus and the transmembrane domain M2 appear to determine whether channel subunits can form tetramers with each other (Tinker *et al.*, 1996). The biochemical and electrophysiological experiments of Tinker *et al.* (1996) suggest that members of the Kir2.0 subfamily are unable to form heteromeric channels with each other (or with Kir1.0 and 6.0). M2 and the proximal C-terminus again confer the incompatability. But the constraint is not universally agreed. Some evidence in the literature points to the possibility of heteromers of Kir2.1 with 2.2 or 2.3. Thus, co-expression of Kir2.1 and 2.3 appears to confer the sensitivity of Kir2.3 to Gβγ onto Kir2.1 (Cohen *et al.*, 1996b). Fink *et al.* (1996) have also proposed that Kir2.1 and 2.3 form heteromeric channels in the nervous system. Gene knockouts of Kir2.1 and 2.2 suggest that both may contribute to some of the ion channels carrying the inward rectifier current I_{K1} in heart (Zaritsky *et al.*, 2001).

2.1.3 Accessory Proteins

Most ion channels have accessory proteins, which do not themselves form pores, but which contribute to the regulation of the channel. In the Kir family, K_{ATP} is formed as a complex of Kir6.0 subunits with the sulphonylurea receptor, SUR (Aguiler-Bryan *et al.*, 1995; Inagaki *et al.*, 1995a, b). Each channel has both four Kir6.0 and four SUR subunits (Clement *et al.*, 1997; Inagaki *et al.*, 1997; Shying & Nichols, 1997). As an ATP-binding cassette (ABC) protein, SUR has two nucleotide binding pockets at the cytoplasmic side of the membrane (Aguiler-Bryan *et al.*, 1995). SUR has the role of regulating the affinity of Kir6.0 for ATP, and of conferring responsiveness to certain drugs, including channel blockage by sulphonylureas and activation by K^+ channel opening drugs. SUR also appears to act as a chaperone protein, essential for normal membrane insertion of K_{ATP} (John *et al.*, 1998; Zerangue *et al.*, 1999), owing to the presence of an ER retention sequence in Kir6.0. It is possible that certain Kir1.0 channels form complexes with the ABC proteins CFTR (McNicholas *et al.* 1996) or SUR (Dong *et al.*, 2001).

Whether other Kir channels require accessory protein subunits is largely unknown. So far β, accessory subunits, common among Kv channels (eg. Rettig *et al.*, 1994), have not been identified for other members of Kir. The study made on Kir2.2 by Raab-Graham & Vandenburg (1998), using cross-linking of subunits, indicates that this channel does *not* possess accessory subunits. Channels such as Kir2.1 do not require chaperone proteins for cell surface expression. Rather channels possess a motif in the C-terminus (residues 374-380, FCYENEV; Ma *et al.*, 2001; Stockklausner *et al.*, 2001) that is necessary for its export from the endoplasmic reticulum and efficient expression in the surface membrane. Mutation of these residues (with the exception of C375) alters cellular trafficking of Kir2.1 after expression in opossum kidney cells from that found in wild type (expression at the surface and in the Golgi complex; Stockklausner *et al.*, 2001). Similarly, Kir3.4 possesses a C-terminal motif that is essential for its transfer and for that of Kir3.1, with which it forms heteromers, to the surface membrane (Kennedy *et al.*, 1999; see section 4).

Several Kir channels (Kir2.1 - 2.3, Kir4.0) bind PDZ domain proteins, which affects their cellular distribution and their ability to act as effector proteins in cell signalling pathways (*e.g.* Cohen *et al.*, 1996a; Hibino *et al.*, 2000). Kir3.2c and Kir3.3 also possess PDZ recognition sequences (Jelacic *et al.*, 2000; Nehring *et al.*, 2000), but their possible binding to PDZ proteins remains controversial (Nehring *et al.*, 2000). We describe properties of anchoring proteins in relationship to Kir2.0 in the next section.

3 Constitutively Active Strong Inward Rectifiers belong to the Kir2.0 subfamily

3.1 Inward Rectifiers Are Activated by Extracellular K^+

Many of the Kir potassium channels that set cellular resting potentials are strong inward rectifiers that are constitutively active in the cell membrane. Most of these channels belong to the Kir2.0 subfamily, though some are formed from subunits of Kir4.0 and 5.0, sometimes - always in the case of Kir5.0 - as heteromers (Pessia *et al.*, 1996; Pearson *et al.*, 1999). Heteromers also form between Kir4.1 or 4.2 and members of the Kir2.0 subfamily (Fakler *et al.*, 1996a; Lagrutta *et al.*, 1996; Pearson *et al.*, 1999). Kir5.1 may negatively control Kir2.1 by forming heteromeric, electrically silent channels (Derst *et al.*, 2001b).

In strong inward rectifiers, the probability of channels being open changes steeply with voltage. The relationship between P_{open} and voltage changes with $[K^+]_o$, shifting along the voltage axis by 25mV for an *e*-fold change in K^+-concentration, giving the dependence on $(V\text{-}E_K)$ first described by Hodgkin & Horowicz (1959b). Adding K^+ to the extracellular solution increases channel P_{open} and the rate of activation at any given membrane potential (Fig. 4). These ion channels are activated by an increase in $[K^+]_o$, and in understanding their function it is sometimes helpful to think of them as *K^+-activated K^+ channels*, a term introduced by Pennefather *et al.* (1992).

One apparently paradoxical result of K^+ activation is that an increase in $[K^+]_o$ may enhance K^+ *efflux* over a range of membrane potentials. This property led to the permeability mechanism first being described as a carrier-mediated mechanism (e.g. Adrian, 1969). However, other groups developed models for the gating of Kir channels where K^+ binding stabilised the channel in the open state (Ciani *et al.*, 1978; Cleemann & Morad, 1979; Oliva *et al.*, 1990; Pennefather *et al.*, 1992). Yet others argued for gating of an ion channel through a simple voltage-dependent ionic blockage from the intracellular milieu (Armstrong, 1969; Hille & Schwarz, 1978; Standen & Stanfield, 1978b). In the days before single channel recording made the matter certain, the argument over carrier versus channel was settled by experiments with blocking cations. Intracellular cationic blockage of the ion channel is widely held to account for channel activation by negative membrane potentials and by elevated $[K^+]_o$. However, channels appear capable of undergoing conformational changes, and the possibility exists that K^+ does stabilise an open channel conformation perhaps through interaction with a site near the outer mouth of the channel as first suggested by Ciani *et al.* (1978; see for example Shieh, 2000).

3.2 Inward Rectifier K+ channels Are Multi-Ion Pores

3.2.1 Voltage Dependent Blockage by Extracellular Foreign Cations

Towards the end of the 1970s, several groups began to show that certain cations could block the inward movement of K^+ and that competition occurred between the extracellular blocking and intracellular permeant cations. Such competition could be most easily explained if there were an ion channel (Armstrong, 1971). Cs^+ blockage has a very steep voltage-dependence, with an equilibrium dissociation constant $K_d(V)$ that changed with voltage according to the equation:

$$K_d(V) = K_d(0).\exp(\delta VF / RT),$$

where $K_d(0)$ is the dissociation constant at 0mV. At its simplest, δ gives the fraction of the voltage field through which Cs^+ moves to its blocking site. Measurements made on egg cells (Hagiwara, Miyazaki & Rosenthal, 1976), in frog skeletal muscle (Gay & Stanfield, 1977), and in cardiac Purkinje fibres (Carmeliet, 1979) all gave values for d between 1.3 and 1.5. This initially unexpected result follows from potassium channels being multi-ion pores (Hodgkin & Keynes, 1955; Hille & Schwarz, 1978), with the change in voltage moving both blocking Cs^+ and permeant K^+ (Hille & Schwarz, 1978). In starfish egg, raising external $[K^+]$ helped trap Cs^+ in the pore, preventing it from returning to the extracellular fluid, a result that also contributes to the high value for δ (Ciani *et al.*, 1978; Ciani *et al.*, 1980). Similar voltage dependent blockages occur with Na^+ (Ohmori, 1978; Standen & Stanfield, 1979), Rb^+ (Standen & Stanfield, 1980) and Ba^{2+} (Hagiwara *et al.*, 1978; Standen & Stanfield, 1978a). Rb^+ blockage is initially increased but is then relieved at more negative voltages, owing to its weak permeance (Standen & Stanfield, 1980). Trace quantities of Ba^{2+} in experimental solutions may sometimes account for a hyperpolarisation-dependent inactivation of Kir2.0 current that is frequently described in the absence of Na^+ (Choe *et al.*, 1998, 1999).

These simple blocking experiments led immediately to mechanisms for dissecting inward rectifier K^+ currents from other currents in voltage clamp studies of a variety of cell types, particularly through the use of Ba^{2+}. Further, when Kir channels were first cloned and expressed, the establishment of blockage by Cs^+ and Ba^{2+} were important diagnostic tests of the nature of the channels (*e.g.* Kubo *et al.*, 1993a; Takahashi *et al*, 1994). Only channels of the Kir7.0 subfamily appear to lack the property of high affinity blockage by Cs^+ and Ba^{2+}, as they lack the strong dependence on external K^+ that is otherwise characteristic of the whole channel family (Krapivinsky *et al.*, 1998b; Döring, *et al.*, 1998).

3.2.2
Measurements of Flux Ratio; Anomalous Mole Fraction Effects

Inward rectifier K^+ channels are multi-ion pores, as is shown by measurements of unidirectional fluxes. The ratio of these is given by:

$$\frac{j_o}{j_i} = \exp\left\{\frac{n(V - E_K)zF}{RT}\right\},$$

where j_o and j_i represent efflux and influx respectively, V is the membrane potential, and E_K is the equilibrium potential for K^+. n gives the minimum number of K^+ ions occupying the channel (Hodgkin & Keynes, 1955). In measurements in skeletal muscle, the value for n was 2 for inward rectifiers (Spalding *et al.*, 1981). Recent measurements in Kir2.1 channels expressed in oocytes give $n = 2.2$ (Stampe et al., 1998), lower, however, than that ($n = 3.4$) found for Kv channels under similar conditions (Stampe & Begenisich, 1996). The role of single filing of ions in achieving maximal conduction rates has been discussed by Morais-Cabral *et al.* (2001).

In addition, inward rectifier potassium channels show anomalous mole fraction effects, where conductance and resting membrane potential is lower in mixtures of permeant cations than in the presence of either ion alone. In mixtures of K^+ and Tl^+, conductance goes through a minimum with about 25% K^+, 75% Tl^+ (Hagiwara *et al.*, 1977; Ashcroft & Stanfield. 1983).

3.3
Gating of Inward Rectifiers

3.3.1
Role of intracellular Mg^{2+}

Experiments with channel blockers contributed the idea that inward rectification could be explained if channels were shut by an internal blocking cation. Armstrong (1966, 1969) first suggested this possibility after showing that intracellular tetraethylammonium ion (TEA^+) induced a form of inward or 'anomalous' rectification in voltage gated K^+ channels of squid axon, with the relief of blockage depending on both voltage and on $[K^+]_o$. The hypothesis was that skeletal muscle metabolises a 'class of TEA^+-like compounds' to produce inward rectification (Armstrong, 1969). The simplest idea for the basis of inward rectification was that an internal blocking cation moved into the channel in a voltage-dependent way so that P_{open} would fall under depolarisation. Occupancy by the blocking cation would be reduced if $[K^+]_o$ were raised, so that at any given voltage, P_{open} would be greater the greater the external K^+ concentration. Hille & Schwarz (1978) made the most complete model of inward rectification on these assumptions. Their use of Eyring rate theory predicted well both the dependence of channel opening on $(V - E_K)$ found experimentally when $[K^+]_o$ is altered and the weaker dependence on $[K^+]_i$ also found experimentally (*e.g.* see Leech & Stanfield, 1981). Their model did not depend on repulsion between ions; rather 'the intimate interdependence of block and the driving force on K^+ ions arises simply from the law of mass action for occupancy' (Hille & Schwarz, 1978).

The blocker considered (Hille & Schwarz, 1978; Standen & Stanfield, 1978b) was monovalent, but the first convincing experimental candidate for a gating particle for inward rectifiers was Mg^{2+}. This suggestion quickly followed the introduction of patch clamp and the use of inside-out membrane patches. Two groups (Matsuda *et al.*, 1987; Vandenberg, 1987) showed that the removal of cytoplasm allowed Kir channels of cardiac muscle to pass large outward K^+ currents. But these currents were blocked by the reintroduction of Mg^{2+} at physiological concentrations (around 1mM). For a period, Mg^{2+} blockage was often seen as sufficient to explain how inward rectifiers were gated.

3.3.2
Mg^{2+} and Substate Behaviour

At low concentrations, Mg^{2+} promoted a sub-state behaviour with unitary currents having amplitudes $^1/_3$, $^2/_3$, and 1 times fully open. It was proposed that this behaviour was caused by the channel having three parallel pores, each independently capable of being occupied by a single Mg^{2+} (Matsuda,

1988). The probability of being at each sub-state fitted this attractive theory. Sub-state behaviour of this kind was also promoted by observations about the various cations that block inward rectifiers from the extracellular fluid (Rb^+, Cs^+, etc; Matsuda *et al.*, 1989).

The triple pore hypothesis is no longer held, at least in its original form. First, transitions between sub-states are not exclusively between adjacent levels ($^1/_3$ to $^2/_3$, $^2/_3$ to fully open, etc) as expected if these transitions represent the binding of a Mg^{2+} (Davies *et al.*, 1996). Secondly, the probability of occupancy of different levels is not appropriate in certain mutant forms of Kir2.1 (Oishi *et al.*, 1998). Though this does not explain the mechanism, the presence of a charged residue appears necessary in the transmembrane domain M2 for the induction of substates (see also Korchev *et al.*, 1997). These transitions disappear in channels in which an asparate residue in M2 has been replaced by one that carries no charge (D172N or D172Q; Oishi et al., 1998). In tandem tetramers where one or two of the four Asp residues present in the channel were replaced by Asn, the $^2/_3$ level is occupied, but not the $^1/_3$. Lu *et al.* (1999a) have recently proposed that the pore is wide enough in the region of D172 to be occupied simultaneously by up to three Mg^{2+} ions and that partial occupancy induces substates.

3.3.3. Gating by Intracellular Polyamines

Mg^{2+} certainly remains a factor of physiological importance in the gating of inward rectifier K^+ channels, but its primacy as the gating particle has been weakened over the past several years. It is known that channels show a slower, time-dependent gating process (Fig. 4). Channel P_{open} increases with time under hyperpolarization and channels shut within a few ms under depolarization (Hagiwara *et al.*, 1976; Hestrin, 1981; Leech & Stanfield, 1981; Kurachi, 1985; Ishihara *et al.*, 1989; Oliva *et al.*, 1990; Silver & DeCoursey, 1990; Stanfield *et al.*, 1994a). Gating depends on voltage and $[K^+]_o$, shifting 25mV per e-fold change in $[K^+]_o$ (Fig. 4). This gating process cannot be due to the release of Mg^{2+} since it persists in the absence of intracellular Mg^{2+} in channels of cardiac myocytes and in Kir2.1 expressed in a cell line (Ishihara *et al.*, 1989; Stanfield *et al.*, 1994a; Fig. 4). Initially the process was thought to be entirely intrinsic to the channel protein, in part because it persisted over long periods under whole cell recording. In Kir2.1 channels expressed in a murine erythroleukaemia (MEL) cell line (Stanfield *et al.*, 1994a), gating was shown to be effectively abolished by replacement of an aspartate residue in the second transmembrane domain by an uncharged residue, Asn or Gln (Stanfield *et al.*, 1994b; Wible *et al.*, 1994).

However, if excised patches of membrane are well washed to remove cytoplasmic constituents - *Xenopus* oocytes seem the most suitable expression system - gating is largely (though not completely) abolished. It can be rein-

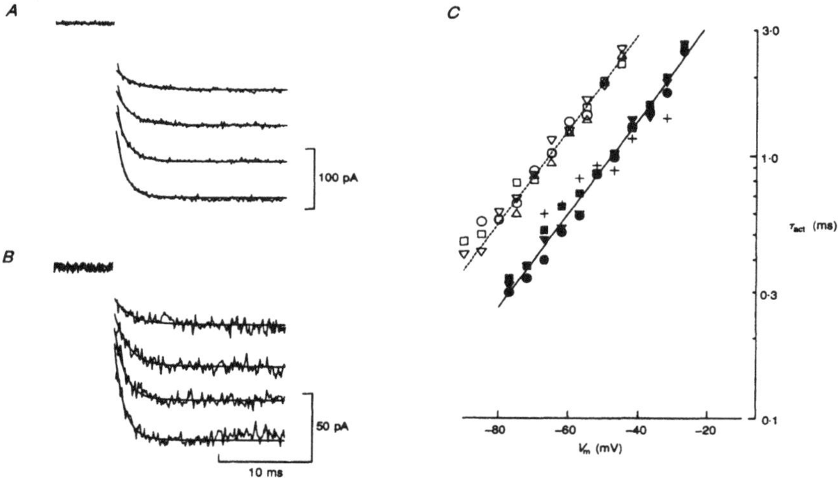

Fig. 4. Activation of inward rectifier K^+ currents. **A, B.** Records of currents obtained under hyperpolarisations of 15, 20, 25, and 30mV from E_K in 70mM (A) and 35 mM $[K^+]_o$ (B). Currents show an initial step increase and then increase along an approximately exponential time course, thought to be due to the release of polyamines, principally spermine. C. Time constants for activation (ordinate, log scale) plotted against membrane potential (abscissa) for 70mM (filled symbols) and 35 mM $[K^+]_o$ in the presence of 1mM $Mg^{2+}{}_i$. +, 70mM $[K^+]_o$ experiments in the absence of internal Mg^{2+}. The continuous line shows that activation time constants for Kir2.1 fall e-fold for a 24mV hyperpolarisation. But time constants are also affect by external $[K^+]_o$; the dashed line is the continuous line shifted 17mV to more negative voltages. Adding K^+ to the external solution increases the rate of activation of ionic currents. Reproduced from Stanfield *et al.* (1994a) by permission.

stated only after provision of a cytoplasmic component other than Mg^{2+}. In the first experiments (Lopatin *et al.*, 1994), patches lost channel gating if they were moved sufficiently far from the oocyte from which they were pulled. But gating was restored if the patch was moved back towards the oocyte, owing to the presence of a substance released from the cell (Lopatin *et al.*, 1994).

Polyamines – putrescine, spermidine and particularly spermine (Fig. 5) – were confirmed by several groups as major gating cations (Lopatin *et al.*, 1994; Ficker *et al.*, 1994; Fakler *et al.*, 1994a; Fakler *et al.*, 1995; Ishihara *et al.*, 1996). Removal allowed the passage of large outward currents, only partly removed by addition of 1mM-Mg^{2+} (Fig. 6). Introduction of spermidine or spermine at low concentrations restored inward rectification (Fig. 6). The gating process, with time dependent reductions in outward currents upon

Ornithine

$$\begin{array}{c} \qquad\qquad\qquad\qquad\quad COOH \\ \qquad\qquad\qquad\qquad\quad | \\ H_2N\cdot CH_2\cdot CH_2\cdot CH_2\cdot CH\cdot NH_2 \end{array}$$

Putrescine

$$H_2N\cdot CH_2\cdot CH_2\cdot CH_2\cdot CH_2\cdot NH_2$$

Spermidine

$$H_2N\cdot CH_2\cdot CH_2\cdot CH_2\cdot NH\cdot CH_2\cdot CH_2\cdot CH_2\cdot CH_2\cdot NH_2$$

Spermine

$$H_2N\cdot CH_2\cdot CH_2\cdot CH_2\cdot NH\cdot CH_2\cdot CH_2\cdot CH_2\cdot CH_2\cdot NH\cdot CH_2\cdot CH_2\cdot CH_2\cdot NH_2$$

Fig. 5. Polyamines involved in gating inward rectifier K^+ channels. Ornithine is converted to putrescine through the action of ornithine decarboxylase; production both first of spermidine and then of spermine is catalysed in two steps by S-adenosylmethionine decarboxylase and aminopropyl transferase.

depolarisation from E_K and time dependent increases under hyperpolarisation, was also restored. Polyamines have less effect on ROMK1 (Kir1.1a; *e.g.* Yang *et al.*, 1995a) or on K_{ATP} (Kir6.0; Niu & Meech, 1998), and these are weak inward rectifiers as a result. Polyamine blockage may be quite widespread among ion channels (Williams, 1997). Extracellular polyamines may potentiate (McGurk *et al.*, 1990), block and permeate glutamate receptors of the NMDA class (*e.g.* Araneda *et al.*, 1999), and weakly block voltage gated Ca^{2+} channels (Sutton *et al.*, 1993). Intracellular polyamines induce inward rectification in Ca^{2+} permeable glutamate receptors of the AMPA/kainate class (Bowie & Mayer, 1995; Isa *et al.*, 1995). They also block both Ca^{2+}-activated K^+ channels (Weiger & Hermann, 1994) and retinal cyclic nucleotide gated channels (Guo & Lu, 2000a), and block and weakly permeate voltage gated Na^+ channels (Huang & Moczydlowski, 2001).

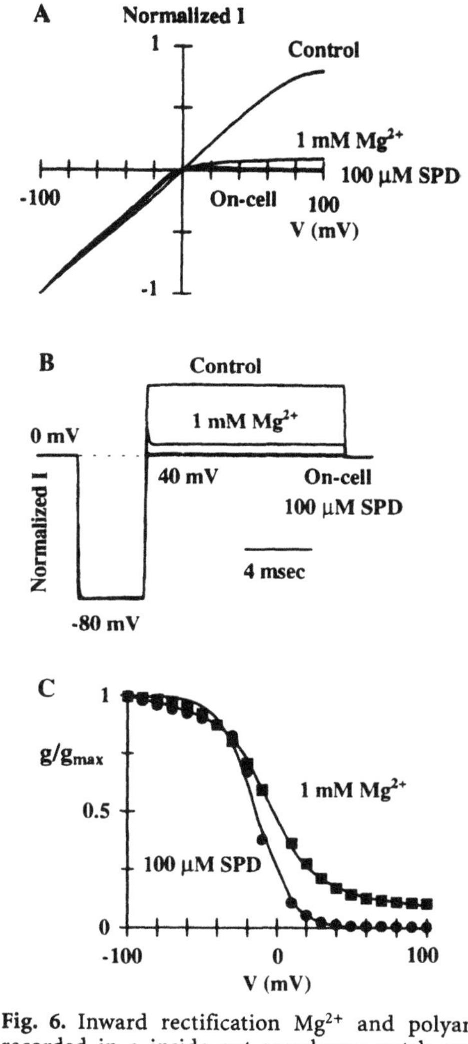

Fig. 6. Inward rectification Mg^{2+} and polyamines. A. Current voltage relations recorded in a inside out membrane patch exposed to solutions without Mg^{2+} or polyamines (control) and in the presence of 1mM-Mg^{2+} or of 100µM spermidine. B. Currents recorded on cell, and then under the same conditions as in A. Note that large outward currents flow in the absence of gating cations. C. Relationship between conductance and membrane potential in the presence of 1mM-Mg^{2+} or of 100µM spermidine. The effective valency of blockage is 1.3 for $Mg2^+$ and 2.1 for spermidine. Reprinted from *Neuron* Volume 14, Yang J, Jan YN & Jan LY, Control of rectification and permeation by residues in two distinct domains in a inward rectifier K^+ channel, page 1049, Copyright (1995), with permission from Elsevier Science.

The polyamines putrescine, spermidine, and spermine are derivatives of ornithine (Fig. 5). Thus blockers of ornithine decarboxylase (ODC), which catalyses the first, rate limiting step in their production, and of S-adenosyl methionine decarboxylase (SAMDC), which catalyses the first of two steps in the formation of both spermidine and spermine, increased outward currents through inward rectifiers. Injection into oocytes of difluoromethylornithine (which irreversibly blocks ODC) and methylglyoxal bis(guanylhydrazone) (which reversibly blocks SAMDC) doubled the expressed Kir2.1 currents and shifted the relation between P_{open} and voltage 10mV more positive (Shyng *et al.*, 1996). In addition, a Chinese hamster ovary cell line (O-CHO) that lacks ornithine decarboxylase gave weakly rectifying currents if deprived of putrescine. In ODC deficient cells, putrescine is required in the media (Shyng *et al.*, 1996), owing to an essential role of polyamines in cell growth and differentiation.

Inhibitors of polyamine synthesis have also been studied in RBL-1 cells, which are derived from a rat basophil leukaemia and which express a strong inward rectifier K^+ channel (Bianchi *et al.* 1996). Here treatment with difluoromethylornithine had little effect on electrophysiological properties. It reduced measured concentrations of putrescine and spermidine, but not of spermine. In contrast, blockage of SAMDC increased outward currents markedly. Here, putrescine was shown to increase in concentration while the levels of spermidine and spermine fell. Since outward currents increased only in these conditions, Bianchi *et al.* (1996) concluded that spermine is the principal blocker of outward currents in the inward rectifier of RBL-1 cells. They also point out that changes in polyamine concentration under the influence of growth factors, mitogens and hormones are a potential route for modulation of Kir channel function (Bianchi *et al.*, 1996). Channel activity might also be expected to change during the cell cycle, as is known to occur for certain potassium channels (Arcangeli *et al.*, 1995).

At physiological pH, putrescine is divalent, with spermidine trivalent and spermine tetravalent since the pK_as of the various amine groups are greater than or much greater than 8. Even if spermine is the principal blocker in RBL-1 cells (Bianchi *et al.*, 1996), all three polyamines are believed to be involved in gating, but the quality of blockage by each is different. The nature of blockage by each polyamine species was studied in most detail in oocytes expressing HRK1 (Kir2.3) channels (Lopatin *et al.*, 1995). Gating of inward rectifiers had been modelled by several workers as a first order process (e.g. Leech & Stanfield, 1981; Ishihara *et al.*, 1989), but such a simple model does not describe channel behaviour completely. For example, the relationship between P_{open} and voltage does not fit a single Boltzmann relationship well, and both activation and the recovery from the shutting of channels under depolarization have instantaneous and slower elements (e.g., Leech & Stanfield, 1981). Lopatin *et al.* (1995) found Kir2.3 currents under hyperpolarization to fit best with a step followed by the sum of two exponentials with time constants τ_f and τ_s. τ_f and τ_s denote faster and slower elements, with τ_f having the weaker voltage dependence (Lopatin *et al.*, 1995). These faster and slower elements were expected to reflect different rates of release of different polyamine species.

When this gating was compared with effects of different polyamines, the following properties were described in Kir2.3. First, blockage by putrescine and its reversal appear instantaneous. Secondly, spermidine and spermine give resolvable increases or decreases in current as blockage is removed or induced. However the relief of blockage by spermidine occurs more quickly than does relief of blockage by spermine. As expected from the lower valency of spermidine, the rate at which it is released from the channel has the weaker voltage dependence. The rate changes *e*-fold every 33mV for spermidine and every 15mV for spermine (Lopatin *et al*, 1995), mirroring the voltage-dependence of τ_f and τ_s respectively as expected. In Kir2.1 also, spermine blockage accounts for time dependent increases in current under hyperpolarisation, whilst blockage by putrescine and Mg^{2+} is virtually instantaneous (Lopatin *et al.*, 1995; Ishihara *et al.*, 1996).

The evidence from Kir2.3 was consistent with two polyamine molecules occupying each channel at positive voltages, just about possible, given that spermine is approximately 2nm in length. Lopatin *et al.* (1995) tested the hypothesis that gating could be first order under conditions where only one polyamine species was present. Channel gating could not be described by the first order model:

$$\textit{Closed} \underset{\beta}{\overset{\alpha}{\longleftrightarrow}} \textit{open}$$

since, even with only, *e.g.*, spermine present, a single Boltzmann does not fit the relation between P_{open} and voltage. Further, the development of blockage by spermidine and spermine has quasi-instantaneous and time dependent elements. The instantaneous element of channel blockage has a shallower voltage-dependence than does the steady state, implying that polyamines may bind initially to a shallow site to block, before moving more deeply into the pore.

The solution given by Lopatin & Nichols (1995) was a kinetic model (Fig. 7) where up to two polyamine molecules block. The first polyamine molecule to enter the channel can move to a deeper blocking site in a transition that is both steeply voltage dependent and is aided by electrostatic repulsion from a second polyamine molecule, if present.

Blockage by polyamines appears to be incomplete at positive voltages (Guo & Lu, 2000b). Even some of the earliest descriptions of inward rectification in skeletal muscle (*e.g.* Adrian & Freygang, 1962b) showed membrane currents that did not fall to zero under depolarisation. This persistent outward current was often treated as leakage or as current through an unidentified K^+ channel and subtracted. But part at least of the persistent outward current

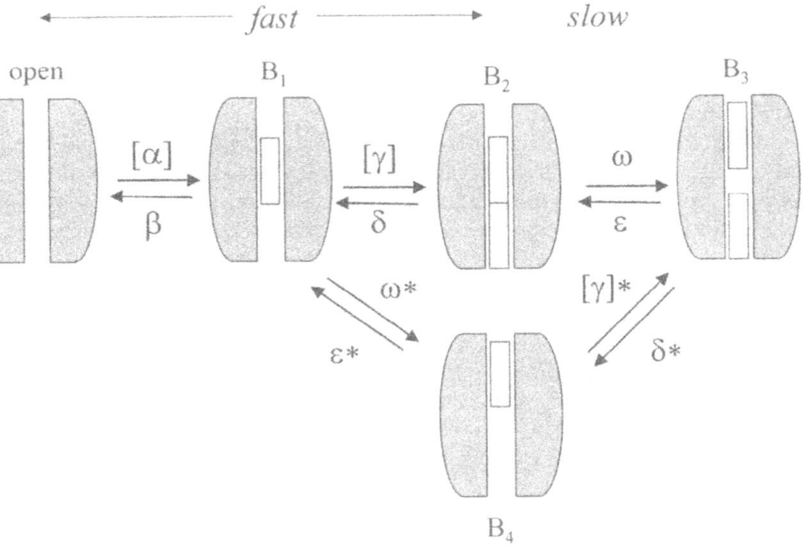

Fig. 7. Scheme for blockage of Kir channels by polyamines. Blockage by polyamines is described as occurring through two or more sites. The initial blocking steps to B_1 and B_2) are rapid and the rate constants α and γ depend on polyamine concentration. The rate constants β and δ depend on membrane potential. An addtional slow, voltage dependent step involves movement of one polyamine molecule deeper into the pore (B_2 to B_3). This movement will be aided by electrostatic repulsion between polyamine molecules. The second slow step may also occur when a channel is blocked by only a single polyamine molecule (B_1 to B_4), but the transition will be governed by different rate constants (ω* and ε*). Drawn after the description given by Lopatin & Nichols (1996a).

may be associated with inward rectifier K^+ channels and with relief of polyamine blockage. Guo & Lu (2000b) compared blockage of Kir2.1 by intracellular TEA^+, putrescine, spermidine and spermine. TEA^+ blockage was voltage dependent and channels became completely blocked at sufficiently positive voltages. But this was not the case with putrescine, where substantial outward currents were seen even at voltages 100 mV positive to E_K. Similarly, spermidine blockage permitted some K^+ efflux at positive voltages. Small outward currents also persisted with spermine.

The simplest explanation was that polyamines can move through the pore at sufficiently positive membrane potentials, relieving blockage (Guo & Lu, 2000b). A comparable phenomenon was found with blockage of inward rectifiers by external Rb^+ and to a lesser extent Cs^+ (native channels, Standen & Stanfield, 1980; Kir2.1, Abrams *et al.*, 1996), which permeate weakly under a sufficient voltage gradient. Polyamines appear to be similarly weakly permeant through Kir channels. The results of Guo & Lu (2000b) imply that blocking polyamines are able to pass through the pore region, H5. Consistent with this hypothesis, argiotoxin, a polyamine toxin that possesses a hydrophobic head group and so would be expected to be completely impermeant, gives complete blockage at positive voltages.

As expected from studies of gating of inward rectifiers in native tissues (Hagiwara & Yoshii, 1979; Hestrin 1981; Leech & Stanfield, 1981) where channel open state probability was found to be principally a function of voltage and $[K^+]_o$, blockage by polyamines is strongly dependent on $[K^+]_o$ (Lopatin & Nichols, 1996a). Only small shifts in the relationship between conductance and membrane potential occur if $[K^+]_i$ is altered.

3.3.4 Interactions of Mg^{2+} and Polyamines

The hypothesis of cationic blockage as the gating mechanism is simple in principle, but the kinetics that result are complex owing to channels having two or more sites of blockage and to the finding that channels may be shut by either Mg^{2+}, putrescine, spermidine or spermine. Blockage by Mg^{2+} and putrescine has low affinity but is instantaneous, whilst that by spermine is of high affinity but occurs in part more slowly. As a result, during a depolarisation channels will be more likely to be shut at short times by Mg^{2+} or putrescine; but at longer times, these blocking cations will be replaced over a period by spermine (Ishihara *et al.*, 1996). Return to a less positive voltage after a short depolarisation will release Mg^{2+} and/or putrescine and the channel will conduct for a while before it is blocked again by spermine (Ishihara, 1997). Ishihara (1997) has shown that substantial transient outward currents may flow as a result (Fig. 8). Ishihara (1997) modelled the phenomenon by supposing that channels may be blocked either by Mg^{2+} or by spermine as follows:

spermine blocked $\longleftrightarrow$ *open* $\longleftrightarrow$ *Mg^{2+} blocked.*

Kir channels are known to contribute to late, phase 3 repolarisation of the cardiac action potential (*e.g.*, Ibarra et al., 1991). The complex kinetics results in Kir2.0 channel contributing a more substantial transient outward current than anticipated from models where inward rectification is treated as instantaneous. Put another way, Mg^{2+} may 'actually enhance Kir current at voltages where the channel is largely intrinsically closed' (Lopatin *et al.*, 1995). Fakler *et al.* (1994a) showed that current-voltage relations generated by ramp changes in membrane potential gave larger (rather than smaller) outward currents in the presence of *both* Mg^{2+} and spermine than in the presence of either blocker alone.

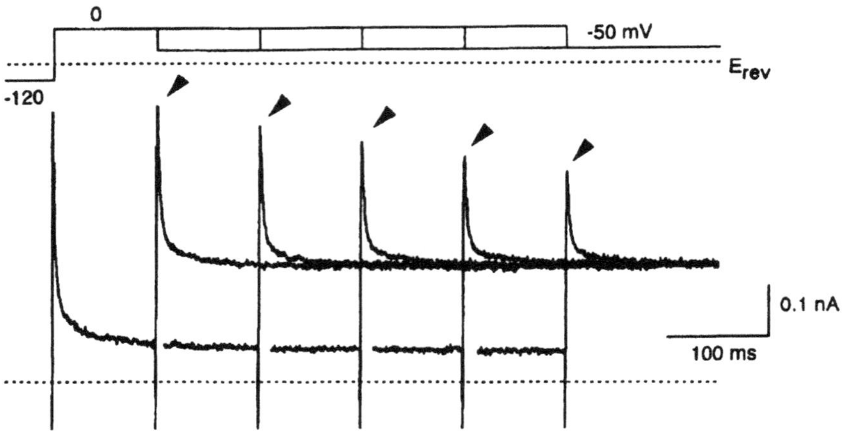

Fig. 8. Outward currents through inward rectifier K^+ channels. Stepping to a depolarised level (0mV) for a brief period and then back to a less depolarised level (-50mV) that is still positive to E_K (E_{rev}) leads to brief, but relatively large outward currents. This transient current is associated with blocking Mg^{2+} and putrescine leaving the channel, which is then blocked again by spermine. Currents become smaller if the depolarisation to omV is prolonged. This is believed to occur owing to the replacement of Mg^{2+} and putrescine as blockers at short times by spermine at longer times. Reproduced from the Journal of General Physiology (1997) 109: 229-243, by copyright permission of The Rockefeller University Press.

3.3.5 Measured Affinities for Blockage by Polyamines.

The affinities of wild type Kir2.1 channels for polyamines have been measured (Ficker *et al.*, 1994; Taglialatela *et al.*, 1994; Yang *et al.*, 1995a). Consistent with the proposal by Lopatin *et al.* (1995) that each channel has more than one blocking site, blockage by any one species of polyamine was found to be the sum of two Hill equations (Yang *et al.*, 1995a):

$$I/I_{con} = I_1\{1+([X]/K_{d1})\}^{-1} + I_2\{1+([X]/K_{d2})\}^{-1}.$$

$[X]$ is the concentration of polyamine producing a fractional current I/I_{con} by binding to sites of high and low affinity whose respective K_ds are K_{d1} and K_{d2}; I_1 and I_2 are the fractions of the current associated with the two blocked states. Thus at high concentration, the pore can accommodate more than one blocking molecule.

At +40mV with symmetrical 140mM [K^+], the measured K_{d1} was 7.5 µM for putrescine, 0.008 µM for spermidine, and 0.0009 µM for spermine (Yang *et al.*, 1995a; Table 3). If the dependence of chord conductance (or channel P_{open}) on voltage is assumed to follow the dependence of affinity on voltage, the effective valency, z', for spermidine was 2.1. The K_d for Mg^{2+} was 17 μM at +40 mV, but its effective valency as a blocking ion was only 1.4.

Guo & Lu (2000b) have also described high and low affinity blockage by spermidine, with two voltage dependent elements, each occurring with a different effective valency. But their explanation for these two modes of blockage is different from that of Yang *et al.* (1995a) and of Lopatin et al (1995). The view of Guo & Lu (2000b) is that different protonation states of spermidine (or possibly of the channel itself) contribute the different blocking affinities. In support of this, reduction of pH_i, to increase protonation, reduced the fraction of blockage that occurred with the lower affinity. Guo & Lu (2000b) obtained a value for z' of ~5 for high affinity blockage by spermidine, higher than that given by Yang *et al.* (1995a). This high effective valency was attributed to the need to move both blocking and K^+ ions within the voltage field (Guo & Lu, 2000b).

3.3.6 Residues that Form the Receptor for Gating Polyamines

If these channels are gated by intracellular cations, it may be anticipated that certain pore lining residues will form a receptor for these molecules. Residues in three positions have so far been shown to be important. These residues are an aspartate residue (D172 in Kir2.1) in the M2 transmembrane domain and glutamate residues (E224 and E299 in Kir2.1) in the C-terminus region (Fig. 2A; Yang et al., 1995a; Taglialatela *et al.*, 1995; Kubo & Murata, 2001). Thus mutation of D172 to Asn or Gln abolishes channel gating around E_K, while mutation to Glu sustains it (Stanfield *et al.*, 1994b; Wible *et al.*, 1994). Gating is introduced into the channel Kir1.1 by replacement of the native Asn (N171) in M2 by Asp. Lu & MacKinnon (1994; see also Wible *et al.*, 1994) argue that this effect is due entirely to the charge on the residue at this position, with Asp & Glu being equivalent, as are Asn and Gln.

These changes were first shown to be associated with changes of affinity for spermine and other polyamines by Fakler *et al.* (1994a), who replaced the native Glu in Kir4.1 with Gln or Asn. Kir4.1 lacks the residues in the C-terminus that have been implicated in polyamine gating (Table 2). Fakler *et al.* (1994a) found that the change E→Q moderately reduced spermine affinity (~15 fold), while the change E→N reduced affinity 100,000 fold, bringing it to the same order as that found for Kir1.1 (Fakler *et al.* 1994). Thus the change in affinity may not be entirely explained by the presence or absence of charge. More detailed measurements have been made by Yang *et al.* (1995a) and by Taglialatela *et al* (1995). In Kir 2.1, the mutant D172N has affinities for spermidine and spermine reduced about 400 fold (Table 3). The affinities for putrescine and Mg^{2+} are less affected, and are reduced 5–10 fold (Yang *et al.*, 1995a). Again similar values are given by Taglialatela *et al.*, (1995).

Table 2. Sequence of domain M2 in Kir channels and in KcsA

Channel name	Sequences of residues (single letter code)	Residue positions	Residue at position 224 or equivalent	Residue at position 299 or equivalent
Kir1.1a	AIFLLIFQSILGVII**N**SFMCGA	156–178	G(223)	D(298)
Kir2.1	AVFMVVFQSIVGCII**D**AFIIGA	157–179	E(224)	E(299)
Kir2.2	AVFMVVAQSIVGCII**D**SFMIGA	158–180	E(225)	E(300)
Kir2.3	AVIAVVVQSIVGCVI**D**SFMIGT	149–171	E(216)	E(292)
Kir2.4	AVAAVVLQCIAGCVL**D**AFVVGA	160–181	E(227)	E(302)
Kir3.1	GIILFLFQSILGSIV**D**AFLIGC	158–180	S(225)	E(300)
Kir3.2	GIILLLIQSVLGSIV**N**AFMVGC	169–191	E(236)	E(311)
Kir3.4	GIILLLVQAILGSIV**N**AFMVGC	164–186	E(231)	E(306)
Kir4.1	AIVLLIAQLVLTTIL**E**IFITGT	143–165	G(210)	S(285)
Kir5.1	AVLTVILQSILSCII**N**TFIIGA	152–174	E(213)	I(283)
Kir6.2	AILILIVQNIVGLMI**N**AIMLGC	145–167	S(212)	E(287)
Kir7.1	AIALLAIQMLLGLML**E**AFITGA	134–156	S(201)	S(272)
KcsA	AVVVMVAGITSFGLVTAALATW	92–113	–	–

Note that in certain channels that are formed as heteromers (Kir3.1 and 3.4 for example & Kir4.1 and 5.1), one only of the channel forming subunits retains the residue (D or E in M2) that acts as the recognition site for polyamines. The subunit retaining this residue lacks the equivalent of E224, which is found in the complementary subunit. Weak rectifiers (Kir1.0, Kir6.0) lack either site. The equivalent of E299 is found in both subunits forming Kir3.0 heteromers but is absent from both Kir4.0 and Kir5.0 and from Kir7.1. Pore lining residues, identified from a homology model with the X-ray structure of the bacterial potassium channel KcsA (Doyle *et al.*, 1998), are shown in italics. Those underlined in Kir2.1were found to be pore-lining by the scanning cysteine mutagenesis studies of Lu *et al* (1999). Minor *et al.* (1999) found S165, C169, D172, and I176 to be pore lining in their mutagenesis studies using a yeast genetic screen. The sequence of KcsA is shown, with an alignment based partly upon identification of pore lining residues in different channels.

Table 3. Blocking affinities for Mg2+ and polyamines in wild type and mutant inward rectifier K+ channels

	Kir1.1	Kir2.1 Wild type		Kir2.1 D172N	Kir2.1 E224G		Kir2.1 E299S	Kir2.1 D172N/ E224G	Kir2.1 E224G/ E299S	reference
		high affinity	low affinity		high affinity	low affinity				
Mg^{2+}	6.0mM	17.1μM	2170μM	82μM	0.12mM	-		0.56mM		Yang *et al.* (1995)
	2.2mM	17μM		30μM	0.18mM			-		Taglialatela *et al.* (1995)*
Putrescine	8.0mM	7.5μM	807μM	66μM	61μM	4.67mM		0.70mM		Yang *et al.* (1995)
Spermidine	3.0mM	8.0nM	2.87μM	3.32μM	0.9μM	0.25mM		0.50mM		Yang *et al.* (1995)
	22.2mM	18mM		0.45μM	6μM			>0.5mM		Taglialatela *et al.* (1995)*
Spermine	1.6mM	0.9nM	0.6μM	0.31μM	1μM	90mM		1.09mM		Yang *et al.* (1995)
		0.5nM			0.2μM		0.1μM		30μM	Kubo & Murata (2001)

K_d values at +40mV for blockage by Mg^{2+} and polyamines of wild type and mutant forms of Kir2.1 are given from the results of Yang *et al.* (1995) and of Taglialatela *et al.* (1995). Both groups used symmetrical K^+ concentrations – 140mM in the case of Yang *et al.* (1995), 100mM in the case of Taglialatela *et al.* (1995). *Certain of the results of Taglialatela *et al* (1995) are cited from other papers of the same group – Taglialatela *et al* (1994; Mg^{2+} affinities for wild type channels); Wible *et al.* (1994; Mg^{2+} affinity of D172N); and Ficker *et al.* (1994; spermidine affinity of wild type channels and of D172N). In both cases, where a single Hill equation was used to fit results, Hill coefficients were between 0.8 and 1.2. Kubo & Murata (2001) conducted their experiments with 20mM $[K^+]_o$ and estimated K_ds at +30mV. Yang *et al.* (1995) measured the voltage dependence of Mg^{2+} and spermidine blockage indirectly from the relationship between K^+ chord conductance and membrane potential, finding effective valencies of 1.37 for Mg^{2+} and 2.13 for spermidine. In I_{KACh} of heart – presumably Kir3.1 and 3.4 heteromers – Yamada & Kurachi (1995) found an IC_{50} for blockage by spermine of 10nM at +40mV.

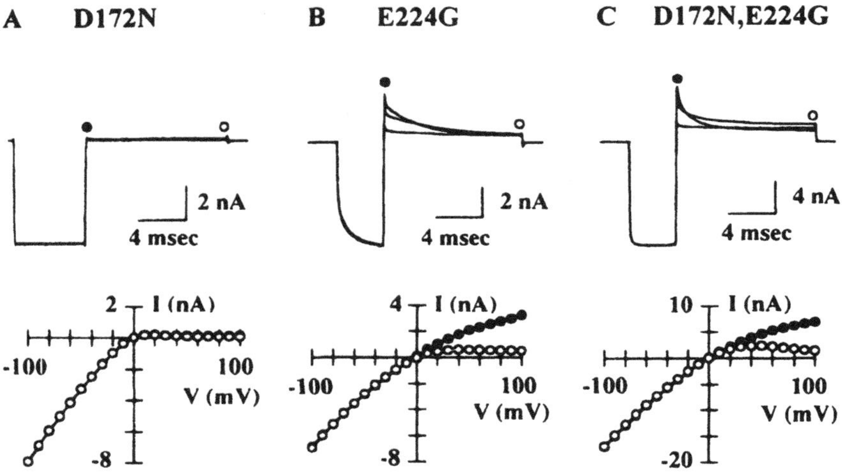

Fig. 9. Currents and current-voltage relations in mutant forms of Kir2.1. Mutants are D172N (A), E224G (B) and D172N, E224G (C). Note that large outward currents are seen in B and C especially at short times under depolarisation. Reprinted from *Neuron* Volume 14, Yang J, Jan YN & Jan LY, Control of rectification and permeation by residues in two distinct domains in a inward rectifier K^+ channel, page 1049, Copyright (1995), with permission from Elsevier Science.

The second residue identified as important in producing strong inward rectification through polyamine binding is E224 in Kir2.1 (Fig. 9). This residue lies in the C-terminus some 45 residues from the cytoplasmic end of M2. The residue is Gly in weak rectifiers such as Kir1.1. Mutations of E224 in Kir2.1 were first carried out by Taglialatela *et al.* (1995) and by Yang *et al.* (1995a). Both groups found reduced affinity for Mg^{2+} and for all gating polyamines in the mutant E224G and in other mutants. Both also described reductions in unitary current and an increase in the current noise in the open level in single channel recording, implying that E224 was a pore lining residue and that the C-terminus was folded to form the inner pore mouth (Taglialatela *et al.*, 1995; Yang *et al.*, 1995a). It is likely that this residue lines the inner vestibule to entry of the channel described by Lu *et al.* (1999b). The effect of mutation of E224 on Mg^{2+} affinity depended on the nature of the substituted side chain. Among the mutations supporting currents, the effects on affinity were in the order E224D (charged) < E224G < E224S < E224Q < E224K (Yang *et al.*, 1995a), with E224Q producing >100 fold reduction in affinity. Effects of these other mutants on polyamine affinity were not investigated by Yang *et al.* (1995a).

Affinities were still more radically reduced by appropriate replacement of both E224 and D172. The two sites appear to act as molecular recognition sites for both Mg^{2+} and for polyamines. The changes in affinity resulting from each mutation and from dual mutations are consistent with the two sites contributing independently but simultaneously to the binding of polyamine blockers (Yang *et al.*, 1995a; see also Wells, 1990). The dual mutation brings affinities close to those found for Kir1.1 (Yang *et al.*, 1995a; Taglialatela *et al.*, 1995), the weak rectifier that lacks both negatively charged residues. But the replacement, in Kir1.1 (ROMK1), of G223 by E had little effect on its inward rectifying properties and did not introduce strong inward rectification (Yang *et al.*, 1995a). Mutation of the equivalent residue in the C-terminus of Kir3.0 (E231Q in Kir3.4) appears to have little effect on channel gating and may have little effect on polyamine affinity in these channels (Lancaster *et al.*, 2000). At the time of the discovery of the roles of D172 and E224 (in Kir2.1), certain other negatively charged residues in the C-terminus were considered as possible candidates for involvement in channel gating (E191, D205 and D249 in Kir2.1). But these were shown to play no role (Taglialatela *et al.*, 1995). More recently however, an additional residue, E299 in Kir2.1, was shown to be important (Kubo & Murata, 2001). This residue was identified by study of the sequence of a weak rectifier sWIRK, a member of the Kir4.0 subfamily discovered in salmon brain. This channel remained a weak rectifier in spite of the presence of a negatively charged residue (Glu) in the transmembrane region M2. Mutation of E299 in Kir2.1 (E299S) has a similar effect to mutation of E224 (E224G). Both mutations lower the affinity for polyamines (see Table 3). Mutation of both residues suggests that the two sites contribute independently to binding.

In addition to those residues so far identified as doing so (D172, E224, E299 in Kir2.1), others may contribute to polyamine gating, if in more subtle ways, since other M2 residues line the pore (Table 2). This seems likely if two polyamine molecules may enter the pore (Lopatin *et al.*, 1995) or if there are both high and low affinity sites (Yang *et al.*, 1995a). The four charges of spermine must be accommodated in the pore. And hydrophobic interactions almost certainly also occur since experiments using mono- or diamines show that affinity is raised by increasing alkyl chain length (Pearson & Nichols, 1998; Guo & Lu, 2000b). The diamines block in a voltage dependent way, but permeate weakly at positive voltages (Pearson & Nichols, 1998; Guo & Lu, 2000b). As already indicated, this property is not quite similar to that shown by quaternary alkylammonium (QA) ion blockage. First, TEA$^+$ does not permeate at positive voltages (Guo & Lu, 2000b; 2001). Secondly, in Kir2.1, but not Kir1.1, internal blockage by TEA$^+$ acts with higher affinity than that shown by longer or shorter chain quaternary ammonium ions (Spassova & Lu, 1998; Guo & Lu, 2001). In Kv, QA blockage operates with a higher affinity if the ethyl group(s) of TEA$^+$ are replaced by groups with longer alkyl chains (Armstrong, 1969; French & Shoukimas, 1981),

Polyamines may traverse the selectivity filter (the H5 or P-region) at positive voltages (Guo & Lu, 2000b). Neveretheless, mutations so far made there do not appear to affect macroscopic gating – and, by implication, gating by polyamines – of Kir2.1 (Guo & Kubo, 1998; So *et al.*, 2001).

3.3.7
Kir Channels May Show Intrinsic Gating

There remains a strong likelihood that some intrinsic gating of the ion channels occurs and that channels can undergo transitions between open and shut states owing to conformation changes of the channel protein. Such changes are unlikely to be comparable to those associated with the opening and closing of Kv or KcsA, or indeed of Kir3.0 (Minor *et al.*, 1999; Yi *et al.*, 2001). However, if channels are opened and shut simply by entry and exit of polyamines (and/or Mg^{2+}), channels would remain open in the absence of any blocking cation. Experiments with channels incorporated into lipid bilayers show that this is not the case (Aleksandrov *et al.*, 1996). Under these conditions, channels open and shut according to approximately first order kinetics, with transition rate constants that are relatively low ($\alpha = 0.3\ s^{-1}$; $\beta = 2\ s^{-1}$) and nearly independent of voltage, giving a low P_{open} of around ~0.15. Further, the unitary current-voltage relation is not linear, but outward currents are smaller than inward, a finding that Aleksandrov *et al* (1996) attribute to a gating process too fast to resolve in the lipid bilayer. They consider that this fast gating is modulated or 'amplified by cytoplasmic Mg^{2+} and/or polyamine binding' to the channel.

Further evidence for intrinsic gating comes from experiments using macropatches of membrane from *Xenopus* oocytes. It was known that some inward rectification and time dependent changes in current remained after long term exposure of macropatches containing inward rectifier K^+ channels to polyamine free solutions. This had been attributed to a failure fully to wash out polyamine, though the effective valency for such blockage may be lower than that by polyamine (Guo & Lu, 2000b). This evidence may not be the strongest if Guo & Lu (2000c) are correct in their assertion that the residual inward rectification is due to the use of HEPES or other zwitterionic buffer, and that it disappears if phosphate buffer is used.

Nonetheless, the evidence of Shieh, John, Lee & Weiss (1996) is as follows. At low concentrations of spermine, raising pH_i from 7.2 to 9.0 slowed channel closure under depolarisation (Shieh *et al.*, 1996). However, in the absence of polyamines, channel closure was accelerated by raising pH_i (Shieh *et al.*, 1996). Such channel closure persisted in the mutant channel D172N, which alters the response to polyamine. Raising pH_i to 9.0 radically accelerated channel closure in this mutant. The gating process was not seen in the weak rectifier ROMK1 (Kir1.1), suggestive of some as yet unidentified element in channel structure that contributes gating only to strong inward rectifiers (Shieh *et al.*, 1996).

The rate at which channels shut under depolarisation showed little dependence on membrane potential, although the amount of current remaining was less at more positive voltages. Shieh *et al.* (1996) postulate a gating process where two open states (O_1 and O_2) are connected by voltage dependent transition rate constants, but where the step to shut (C) is independent of voltage, but dependent on pH_i. They were able to fit their results if channels move between the following states in the absence of polyamines.

$$O_1 \underset{\lambda(V)}{\overset{\mu(V)}{\longleftrightarrow}} O_2 \underset{\alpha}{\overset{\beta}{\longleftrightarrow}} C$$

where $\mu(V)$ and $\lambda(V)$ are voltage dependent and α and β are independent of voltage. The rate constant β was substantially increased at alkaline pH_i. The step between the two open states had a voltage dependence equivalent to the movement of approximately 4 charges. The suggestion again was that Mg^{2+} and polyamines enhance the efficacy of this apparently intrinsic kinetic process.

In addition to these findings in wild type Kir2.1 channels, mutations at position 224 in Kir2.1 – mutations that reduce polyamine affinity so radically – also introduce a gating process that is independent of Mg^{2+} or polyamine (Taglialatela *et al.*, 1995; Yang *et al.*, 1995a). Here slow reductions in current occur under depolarisation from E_K and slow increases occur under hyperpolarisation (Taglialatela *et al.*, 1995; Yang *et al.*, 1995a). The kinetics and the degree of inward rectification produced in the steady state depend on the nature of the residue introduced at position 224. Thus inward rectification is strongest if a basic residue is used (K, R, H), is only slightly weaker with the polar residues N & Q, is weaker still with S and then with G, and, apart from the situation in wild type, is weakest in E224D. The introduction of a positively charged residue in place of D172 also induces a rectification that appears to be independent of polyamine blockage with current transients whose rates are not strongly dependent on voltage (So *et al.*, 2001). Kubo & Murata (2001) have also shown persistence of gating and inward rectification in a triple mutant of Kir2.l (D172N/E224G/E299S) that has negligible affinity for polyamine; this rectification is absent from the weak rectifier sWIRK under similar conditions.

The hypotheses of intrinsic gating and polyamine blockage have then often been treated as separate. It is clear first that Mg^{2+} and polyamines are necessary for normal physiological behaviour of Kir2.0 channels. But secondly, if Mg^{2+} and polyamines are absent or if their affinity is markedly reduced by mutation, channels show gating and this gating generates a weak inward rectification. Thus the possibility exists, as several authors have suggested (Aleksandrov *et al.*, 1996; Lee *et al.*, 1999) that Mg^{2+} and polyamines act as co-factors to set the normal physiological properties of an intrinsic gating process.

Lee, John & Weiss (1999) have proposed a mechanism of inward rectification that links the two hypotheses (Fig. 10). The mechanism is somewhat

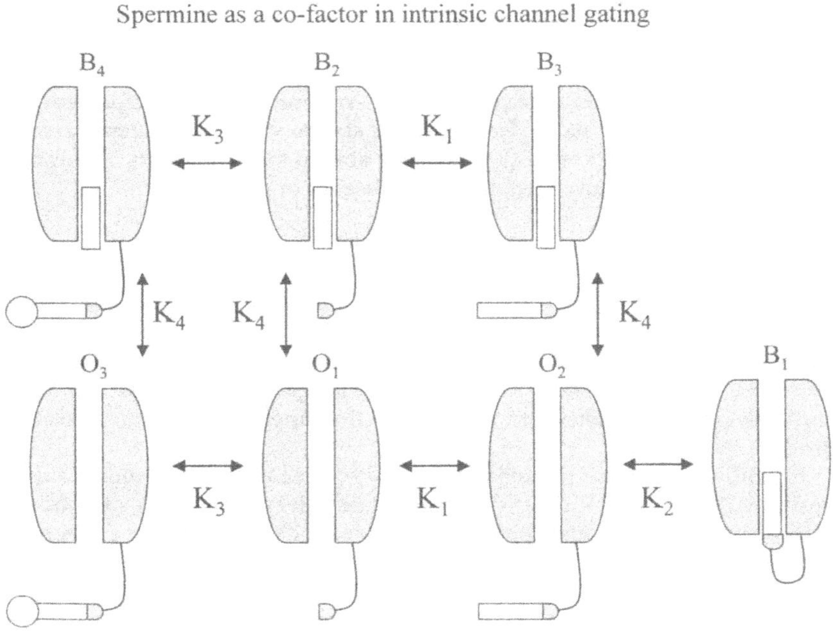

Fig. 10. Scheme for gating that is intrinsic, with spermine acting as cofactor. High affinity blockage involves spermine binding to the C-termiunus, which then moves to lug the channel (O_1 to O_2 to B_1); effectively spermine potentiates an intrinsic gating process. This gating is prevented of polyamine toxin binds to the C-terminus in place of spermine (O_1 to O_3). Spermine can also give simple channel block through a low affinity interaction (B_2, B_3 and B_4). Steps where polyamine toxins block channels through low affinity in the pore are omitted. Drawn after the description by Lee *et al.* (1999).

akin to the N-type inactivation of Kv channels, which occurs through a ball and chain mechanism (Hoshi *et al.*, 1990). In Kv the N-terminus of the channel α– or β-subunit moves to plug the inner part of the pore of certain Kv channels (see also Rettig *et al.*, 1994; Zhou *et al.*, 2001). In Kir2.1, Lee *et al.* (1999) studied effects of the polyamine toxins philanthotoxin and argiotoxin. These toxins are similar in structure to spermidine and spermine respectively, but carry a large hydrophobic head group in place of one of the terminal $-N^+H_3$ groups. The toxins block Kir2.1 but with low affinity, suggesting that simple pore block is possible. Blockage is likely to occur with the polyamine toxin molecule inserted lengthways into the pore as Lopatin *et al.* (1995) propose; the toxin hydrophobic head group will not enter the pore but the hydrophilic tail will. However such a process accounts only for low affinity blockage (see Yang *et al.*, 1995a).

The principal effect of the toxins is to reduce substantially the affinity for spermine and spermidine, primarily affecting the high affinity blockage of Yang *et al.* (1995a). Thus nM concentrations of toxin reduce affinity for the polyamines some 10-fold, a much greater reduction than would be expected if there were simple competition for the pore-occluding blocking site. This effect is lost in the mutant E224G, though low affinity blockage by the toxin is unaltered.

One explanation is that the toxin binds to a site on the C-terminus at or close to E224 through its charged, hydrophilic tail. But the toxin does not block when bound there since its hydrophobic head group prevents its insertion into the pore. However such binding prevents spermine (or spermidine) from binding to E224, through which its high affinity blockage is exerted. Since the site on the C-terminus may be occupied without channel blockage, the hypothesis is that the positively charged polyamine confers voltage dependence on this element of the C-terminus, which then acts as a gate, shutting under depolarisation. By analogy with what is now known to happen in N-type inactivation (Zhou *et al.*, 2001), binding of polyamine first to the C-terminus might generate a pre-inactivated, but conducting state, that leads rapidly to a closed state. Occupancy of the closed state is opposed by channel occupancy by K^+.

This attractive hypothesis is vulnerable to criticism in its present form. First, Yang *et al.* (1995a) had evidence for simultaneous and independent binding at D172 and E224, whereas the hypothesis of Lee *et al.* (1999) envisages binding at D172 being at least partly dependent on prior binding at E224 or thereabouts. Secondly, modification of E224 leads to changes in unitary conductance and open channel current noise (Taglialatela *et al.*, 1995; Yang *et al.*, 1995a) so that E224 is likely to be a pore lining residue, rather than outside the conducting pore. Thirdly, the residue E224 is at the centre of a domain of the C-terminus that helps bind phosphatidylinositol 4,5-bisphosphate (Zhang *et al.*, 1999; see below). Such binding is essential for channel constitutive activity. The hypothesis remains attractive as an explanation for many of the phenomena that relate to Kir2.0, but may need modification in detail.

3.3.8 Comparison with N-Type Inactivation in Kv

A comparison of the gating of Kir2.0 with N-type inactivation of Kv channels may seem far-fetched at first. Certainly the N-terminus of Kir2.0 is not involved since removal of the bulk of this cytoplasmic domain (up to M0) is without effect on gating of Kir2.1 (Shelton *et al.*, 1995). One member of the Kir family at least (K_{ATP} in skeletal muscle) is not readily blocked by ShB ball peptide (Beirão *et al.*, 1994), the inactivating N-terminal peptide from *Shaker* potassium channels (Hoshi *et al.*, 1990).But there are similarities nonetheless. First it is now established in N-type inactivation of Kv that the N-terminus moves into the pore in extended, linearised form (Zhou *et al.*, 2001), binding first to give a pre-inactivated, but open state, and then moving rapidly into the pore to block. The pre-inactivated state may be compared with the first step in high affinity binding of polyamine proposed by Lee *et al.* (1999), the step at which polyamine toxins block gating. Secondly, the site of blockage is the same – or partly the same (Guo & Lu, 2001) – as that occupied by quaternary ammonium compounds such as TEA^+ (Choi, Aldrich & Yellen, 1991) and tetrabutylammonium (TBA^+) ions or tetrabutylantimony (Zhou *et al.*, 2001). Polyamines are the TEA-like compounds proposed by Armstrong (1969) as generating inward rectification, for example in muscle. Thirdly, external K^+ reduces occupancy of channels by the inactivation gate in Kv and raising $[K^+]_o$ accelerates recovery from inactivation at a given negative membrane potential (Demo & Yellen, 1991). The effect of $[K^+]_o$ seems less than that found in Kir channels (where rates shift 25mV *per e*-fold change in $[K^+]_o$), but is still substantial (Demo & Yellen, 1991, their Fig. 2). Finally the residues in S6 of Kv1.4 and in KcsA that act as receptors for the inactivation gate and for TBA^+ respectively include the residue equivalent to D172 in Kir2.1. The residues in Kv1.4 are V551, V554, V558, V562 and Y569 for the receptor for inactiving ball peptide and V551, V554, V558, and V562 for TBA^+ (Zhou *et al.*, 2001). V558 is equivalent to T107 in KcsA (Table 3.1) and to D172 in Kir2.1. The two mechanisms of inactivating ion channels may have evolved in parallel; certainly they use some common structural elements.

3.4 Constitutive Activity Requires Phosphatidylinositol 4,5-Bisphosphate

Several groups have found that Kir2.0 channels run down in whole cell or excised patch recording and that this run down may be reversed by exposure to ATP. In lipid bilayers, Kir2.1 channels have a low P_{open} (~0.15; Aleksandrov *et al.*, 1996). Recently it has been shown that Kir2.1 channels require phosphatidylinositol 4,5-bisphosphate (PIP_2) in the cell membrane. This requirement is characteristic of the whole Kir family for their normal physiological properties. Thus PIP_2 is necessary for a number of physiological responses of Kir channels. First it is needed for constitutive activity in the case of Kir1.0 and Kir2.0 (Liou *et al.*, 1999; Zhang *et al.*, 1999), with PIP_2 interaction with Kir1.1a being enhanced by protein kinase A phosphorylation (Liou *et al.*, 1999). Secondly, it is required for normal activation of Kir3.0 by G-protein $\beta\gamma$ subunits and by Na^+ (Huang *et al.*, 1998; Zhang *et al.*, 1999; Ho & Murrell-Lagnado, 1999a, b). Thirdly, it regulates open state probability and sensitivity to protons of Kir4.1/5.1 heteromers (Yang *et al.*, 2000). Finally, it regulates the ATP-affinity of Kir6.0 (K_{ATP}) with increasing concentrations of PIP_2 raising channel P_{open}, apparently displacing ATP from its binding site on Kir6.0 subunits (K_{ATP}; Hilgemann & Ball, 1996; Fan & Makielski, 1997; Shyng & Nichols,

1998, 2000). PIP_2 is an important cofactor also in the function of other membrane transport proteins, inhibiting both rod cyclic nucleotide gated channels (Womack *et al.*, 2000) and the capsaicin receptor VR1 (Chuang *et al.*, 2001). It is required for activity of the Na^+/Ca^{2+} exchanger in the surface membrane (Hilgemann & Ball, 1996) and the inositol trisphosphate (IP_3) receptor in endoplasmic reticulum (Zhainazarov & Ache, 1999). Phosphoinositides are also important in promoting actin polymerisation and cytoskeleton-membrane interaction (Janmey *et al.*, 1999).

The binding of PIP_2 appears to depend on basic amino acid residues in the C-terminus. First identified was a PKKR motif that is found in the proximal C-terminus, just beyond M2 (*e.g.* Huang *et al.*, 1998; Soom *et al.*, 2001; see Janmey *et al.*, 1999 for a summary of phosphoinositide binding sites in other, cytoskeletal proteins). These residues are at positions 186-189 in Kir2.1. Mutation within this motif affects the action of PIP_2 in most Kir channels (Huang *et al.*, 1998; Baukrowitz *et al.*, 1998; Yang *et al.*, 2000). In Kir2.1, mutation reduces both open state probability and binding of PIP_2 by the C-terminus (Soom *et al.*, 2001). The effect on reducing channel P_{open} was found to be in the order K188Q/R189Q > K188Q > wild type. Mutation of Pro186 (P186A) also reduced P_{open} somewhat. Effects on PIP_2 binding were assessed by measuring binding of a GST (glutathione-*S*-transferase) fusion protein to liposomes containing PIP_2, using binding of GST alone and a GST-phosphatidylinositide 3-kinase γ (GST-PI3Kγ) fusion protein to calibrate the fluorescence method used in measurement. Binding was reduced in the order K188Q/R189Q > P186A, K188Q > R189Q > wild type, not quite following the order found for the reduction in P_{open}. P186A permits formation of functional channels (see above) yet reduces PIP_2 binding substantially, suggestive of some unknown complexity in the control of P_{open} by PIP_2.
A more distal region of the C-terminus of Kir2.1 is also essential. This region lies between residues 324 and 365 in Kir2.1, with the double mutant R343Q/K346Q showing both very low channel activity (P_{open}) and PIP_2 binding (Soom *et al.*, 2001).

While the PKKR motif in the proximal part of the C-terminus is crucial, this motif is present in most members of the channel family, being replaced only in Kir2.3 by SKKR, in Kir3.3 by PNKR, in Kir5.1 by NRKR, in Kir6.0 by AHRR, and in Kir7.1 by PKNR. Yet, constitutively active K^+ channels (e.g., Kir1.0, 2.0) bind PIP_2 more tightly than do Kir3.0 (Zhang *et al.*, 1999). Antibody to PIP_2, applied to inside-out macropatches containing expressed channels, causes loss of channel activity. The rate at which this occurred was used to assess the tightness of PIP_2 binding. Antibody switches off Kir2.1 currents over a period of few minutes (Zhang *et al.*, 1999) when applied at appropriate concentration (Liou *et al.*, 1999). Antibody also switches off a constitutively active mutant of Kir3.4 (S143T), a channel that would, in wild type, require Gβγ to be active, over a few s. Three residues in the C-terminus are principally important in generating this difference in binding between Kir2.1 and

Kir3.4. Mutations R218Q and R228Q much enhanced the rapidity with which antibody caused loss of ionic current through Kir2.1. So did mutation L222I, for reasons that were not clear. The sequence of Kir2.1 around this region is G**N**L**R**KSH**L**V*E*AHV**R**AQ (215-230). Mutation of Asn216 to Asp (N216D) conferred Na^+ sensitivity on Kir2.1 (Zhang *et al.*, 1999; Ho & Murrell-Lagnado, 1999). Na^+ is believed to enhance PIP_2 binding by masking the negative charge introduced at position 216 in the mutant channel. This part of the channel, determining the affinity for PIP_2, surrounds one of the residues (E224, italicised in the sequence above) that is central to gating by polyamines.

As well as the difference in affinity for PIP_2 between Kir2.0 and Kir3.0, there are differences in specificities for different phosphoinositides. Kir2.1 shows a more specific requirement for the lipid hydrophilic head group, while Kir3.1/3.4 heteromers have a more specific requirement for the hydrophobic acyl chains (Rohács *et al.*, 1999). Kir 3.1/3.4 is activated with equal efficacy by phosphatidylinositol 4,5-bisphosphate, PI-3,4-bisphosphate, PI-3,5-bisphosphate, and by PI-3,4,5-trisphosphate. But Kir2.1 is best activated by phosphatidylinositol 4,5-bisphosphate, while it is activated with low efficacy only by high concentrations of PI-3,5-bisphosphate or of PI-3,4,5-trisphosphate. It is not activated by PI-3,4-bisphosphate (Rohács *et al.*, 1999). However, while Kir3.1/3.4 is best activated by the naturally occurring arachidonyl stearyl PIP_2, Kir2.1 has less stringent requirements of the acyl chains and is equally well activated by dipalmitoyl PIP_2. Both require membrane incorporation of PIP_2 for activation, since water-soluble analogues are without effect (Rohács *et al.*, 1999).

Thus, constitutive activity crucially depends on the binding of PIP_2 to Kir2.0 channels. This property offers two mechanisms by which channels may be down-regulated by cell signalling mechanisms through depletion from the membrane of PIP_2. This depletion may occur either through PIP_2 breakdown by phospholipase C or through its phosphorylation by PI3 kinase to phosphatidylinositol 3,4,5-trisphosphate (PIP_3). The second mechanism may be ineffective where channels are themselves activated by PIP_3, as is the case with Kir3.0 (Rohács *et al.*, 1999). So sensitive is Kir2.1 channel activity to PIP_2 levels, that Kir2.1 has been used as a biosensor for PIP_2 levels in *Drosophila* photoreceptor membranes (Hardie *et al.*, 2001) into which it had been inserted by appropriate genetic manipulation (Baines *et al.*, 2001).

3.5 Dependence on $[K^+]_o$

In Kir channels, increasing $[K^+]_o$ increases P_{open}. This phenomenon, where external, permeant K^+ competes with a gating or blocking cation coming from inside, is expected. But in Kir channels, an *e*-fold increase in $[K^+]_o$ is equivalent to shifting the membrane potential to a value 25 mV more negative – that is channel P_{open} is precisely dependent on $(V\text{-}E_K)$. Though models have been able to reproduce this behaviour well (Ciani *et al.*, 1978; Hille & Schwarz, 1978; Standen & Stanfield, 1978b; Cleemann & Morad, 1979; Oliva *et al.*, 1990; Pennefather *et al.*, 1992), it remains unclear exactly what part of the structure is responsible for such precise matching of behaviour to changes in $[K^+]_o$.

One clue comes from a difference in the behaviour of Kir7.1 from that of other members of the Kir family. Kir7.1 lacks, to a great extent, the property of being activated by K^+_o (Döring *et al.*, 1998, see their Fig. 5; Krapivinsky *et al.*, 1998b). Inward rectification is very weak and outward currents are nearly as large as inward currents at an equivalent driving force. Unitary conductance is very low (50 fS in symmetrical 150 mM-K^+_o; Krapivinsy *et al.*, 1998b). Macroscopic conductance is nearly independent of $[K^+]_o$ (Döring *et al.*, 1998b), in contrast to most other inward rectifiers (Hagiwara & Yoshii, 1979; Lopatin & Nichols, 1996b). In most inward rectifiers, conductance is proportional to $[K^+]_o^{0.5}$ (Hagiwara & Yoshii, 1979; Lopatin & Nichols, 1996b). The crucial difference appears to lie in the structure of the outer mouth of the channel selectivity filter (Table 4). For, while all other members of the Kir channel family have an Arg residue immediately at the top of the selectivity filter - the sequence in Kir2.1 is `ETQTTIGYGFR` (138–148) - uniquely in Kir7.1, this residue is replaced by Met (sequence is `ETQLTIGYGTM`, 116–126; Krapivinsky *et al.*, 1998b).

Yang *et al.* (1997) had proposed that in Kir2.1 R148 forms a salt bridge with E138 in the immediately adjacent subunit. This salt bridge would help stabilise the selectivity filter. Most mutations of either residue (R→K, H, C, or E; E→D, Q, C, or R) result in loss of function in channels, though the substitutions E138Q and R148H could be made in two of the four linked subunits in tandem tetrameric constructs. The double mutation E138R/R148E produced functional channels that were, however, permeable to both K^+ and Na^+. P_{Na}/P_K increased from 0.02 in wild type to 0.91 in the double mutant (Yang *et al.*, 1997), indicating a subtle change in the structure of the selectivity filter of the still functional channel. The pH sensitivity of channels - reduction in activity occurring only at very low pH_o - was consistent with the hypothesis that protonation of E138 would disrupt a salt bridge (Yang *et al.*, 1997).

But Kir7.1 must function without such a structure, though its unitary conductance is increased nearly 20-fold by replacing its native Met by Arg (M125R in Kir7.1; Krapivinsky *et al.*, 1998b). Its sensitivity to Ba^{2+} as a blocking cation is also increased (Krapivinsky *et al.*, 1998b; Döring *et al.*, 1998). Inward rectification is strengthened somewhat and outward currents become small (Döring *et al.*, 1998). Further, macroscopic conductance now becomes strongly dependent on $[K^+]_o$, increasing with $[K^+]_o^{0.8}$ (Döring *et al.*, 1998; see also Hagiwara & Yoshii, 1979; Lopatin & Nichols, 1996b). Thus the presence of an Arg residue at the outer mouth of the filter strengthens the K^+ dependence of this inward rectifier.

Other work has shown that some substitutions of R148 can be made in Kir2.1. One such mutant is R148Y (Kubo, 1996). Here activation is slower and

Table 4. The selectivity filter of some K+ channels.

KcsA	TYPRALWWSVETATTVGYGDLY	61– 82
Kir1.1a	GMTSAFLFSLETQVTIGYGFRF	127–148
Kir2.1	SFTAAFLFSIETQTTIGYGFRC	128–149
Kir2.2	GFMAAFLFSIETQTTIGYGLRC	129–150
Kir2.3	GFLGAFLFSVETQTTIGYGFRC	120–141
Kir2.4	SFLAAFLFALETQTSIGYGVRS	131–152
Kir3.1	NFPSAFLFFIETEATIGYGYRY	129–150
Kir3.2	GFVSAFLFSIETETTIGYGYRV	140–161
Kir3.4	GFVSAFLFSIETETTIGYGFRV	135–156
Kir4.1	TLTGAFLFSLESQTTIGYGYRY	114–135
Kir5.1	SFTAAFLFSLETQTTIGYGYRC	117–138
Kir6.2	SFSSAFLFSIEVQVTIGFGGRM	116–137
Kir7.1	SFTSAFSFSLETQLTIGYGTMF	105–126

The pore helix, found through the solution of the X-ray crystallographic structure of KcsA (Doyle *et al.*, 1998), is shown in italics. The residues whose peptide bonds provide the backbone carbonyl oxygens that confer selectivity are underlined in the KcsA sequence (Doyle *et al.*, 1998). The pore helix of Kir channel lacks the Trp residues found in KcsA (67, 68) thought to be involved in stabilising the dimensions of the selectivity filter though a H-bond with Tyr of GYG. Unusually among K^+ channels that form as tetramers, the heteromeric channel formed from Kir3.1 and 3.4 has P-regions that are not identical. This lack of four fold symmetry appears necessary for high K^+ selectivity in this case (Silverman *et al.*, 1998). Among the inward rectifier family, Kir6.0 has Phe rather than Tyr in the K^+ channel signature sequence, a property shared with channels of the *eag* gene family (Warmke & Ganetsky, 1994). The penultimate residue indicated is Arg throughout the *Kir* gene family with the exception of Kir7.1, which lacks the strict (V-E_K) dependence otherwise characteristic of the family. This Arg residue also helps to confer rapid transfer of K^+ and high affinity blockage by Cs^+ & Ba^{2+} (Krapivinsky *et al.*, 1998).

has a much altered K^+ dependence, gating parameters no longer shifting 25mV for an *e*-fold change in $[K^+]_o$. Again, this is consistent with a central role for this basic residue in ionic regulation – that is in activation of Kir channels by external K^+. Further evidence for the role of R148 comes from work by Shieh *et al.* (1999) who successfully expressed the Kir2.1 mutant R148H. In this mutant, unitary conductance became dependent on pH_o (but not on pH_i) being highest when the His residue was protonated (Shieh *et al.*, 1999). The effect of changing pH_o was independent of voltage, indicating that it occurs through a protonation at the outer mouth of the pore. Open channel noise was highest when the pH was close to the pK_a for the effect on unitary conductance. The noise then probably occurs because of high frequency fluctuations between protonated and unprotonated forms of His residues at position 148. The effect of $[K^+]$ on unitary conductance, studied in symmetrical K^+_o/K^+_i, showed that the affinity for K^+ and the channel had become dependent on pH_o, with the affinity being raised with reduced protonation of His. In wild-type Kir2.1 channels the relationship between γ and $[K^+]$ may be fit with a Hill equation with a K_d of 49mM. In R148H at pH 6.0, the K_d is 30mM, while at pH7.4, it appears to be less than 10mM. Thus a positively charged residue at position 148 results in lower affinity for K^+ but faster transfer. Shieh *et al.* (1999) propose that R148 and its positive charge contribute an essential element in a series of K^+ binding sites enhancing electrostatic repulsion between K^+ ions in a multi-ion pore (see above). This suggestion is consistent with the finding that in Kir7.1 unitary conductance is raised if the equivalent Met residue (M126) is replaced by the Arg found in all other inward rectifiers (Döring *et al.*, 1998; Krapivinsky *et al.*, 1998b). Taken together the evidence points to the positively charged Arg residue at the outer mouth of the pore contributing to a K^+ co-ordination site responsible for conferring the dependence on $(V - E_K)$ shown by Kir channels.

Shieh (2000) has also shown Kir2.1 channels to possess a voltage-dependent inactivation mechanism against which K^+ has a protective action. Channels move to a shut state at negative voltages over a period of a few 100ms, but the binding of K^+ prevents channel closure and keeps channels open. The experiments were performed using macropatches from oocytes in which channels were expressed. Increasing $[K^+]_o$, either under conditions of symmetrical $[K^+]$ or at constant $[K^+]_i$, reduced inactivation. This effect was independent of membrane potential over the range −100 to −200mV. It could be fitted with a Hill equation with a K_d of about 5mM, an affinity higher however than that found for the effect on unitary conductance (K_d ~ 50mM). The Hill coefficient was close to 1, so the binding of a single K^+ at a site at the outer mouth of the pore seems possible. The scheme might be presented

open-K^+	$\longleftrightarrow$	*open*-	$\longleftrightarrow$	*closed*-
K^+-bound	*K^+-dependent*	K^+-unbound	*voltage-dependent*	K^+-unbound

In agreement with this hypothesis, mutation of Arg 148 to Tyr much enhanced the potency of K^+ in its protective effect against inactivation, consistent with removal of the positive charge enhancing affinity for K^+. Since the mutant R148Y reduced K^+ flux through the channel, the effect is not dependent on the rate of K^+ transfer. Other permeant cations appear not to be able to replace K^+ since currents are more likely to inactivate when carried by Rb^+ (Reuveny *et al.*, 1996). Some permeant cations may induce inactivation. In native inward rectifiers in frog skeletal muscle, currents carried by Tl^+ inactivate in a voltage dependent way, while those carried by K^+ (under conditions of high $[K^+]$) do not (Stanfield *et al.*, 1981; Ashcroft & Stanfield, 1983). Recently Shieh & Lee (2001) have shown that NH_4^+ and Tl^+ induce voltage dependent inactivation in Kir2.1.

If there is a K$^+$ co-ordination site at this point, the pore may be expected to be narrow there. Evidence that this is so comes from the action of sulfhydryl reagents on Kir2.1. Ag$^+$ covalently binds to the –SH group on cysteine. A single Ag$^+$ is sufficient to block a single Kir2.1 channel (Dart *et al.*, 1998a) through an action on C149, the residue immediately next to R148 (see Table 4). Since Ag$^+$ has approximately the same ionic radius as K$^+$, the pore seems likely to be the correct size for co-ordination with K$^+$ at this point. This Cys residue at which Ag$^+$ blocks is in the equivalent position to that in Kv conferring a K$^+$ regulation site for availability of K$^+$ channels (Pardo *et al.*, 1992) or a site regulating C-type inactivation (Lopez-Barneo *et al.*, (1993) as well as the site for external TEA$^+$ blockage (MacKinnonn & Yellen, 1990). Shieh (2000) has compared the inactivation process in Kir2.1, prevented by K$^+$ binding, with C-type inactivation in Kv.

In Kir1.0 and Kir4.0 at least, K$^+$ is also involved in stabilising channels in an open conformation, though the mechanism is not identical. Channels are non-conducting in the absence of external K$^+$, but become conducting if K$^+$ is introduced into the extracellular solution. This transition occurs quite slowly, over a period of several s (Giebisch, 1998; Doi *et al.*, 1996; Pearson *et al.*, 1999). In Kir1.1, the transition is opposed by protonation, which also gates these channels. This gating is central to the physiological role of Kir1.0 in the secretion of K$^+$ in renal collecting duct and the effect of acidosis or alkalosis upon K$^+$ secretion (Giebisch, 1998). In experimental conditions, and unlike the situation with Kir2.1 (Shieh, 2000), NH_4^+, Rb$^+$ or Cs$^+$, but not Li$^+$ or Na$^+$, can replace K$^+$ in channel activation (Doi *et al.*, 1996; Schulte *et al.*, 2001). Two transitions are believed to be linked (Schulte *et al.*, 2001) as follows:

open-K$^+$	$\longleftrightarrow$	*closed*-K$^+$	$\longleftrightarrow$	*closed*-
unprotonated,		Protonated,		Protonated,
K$^+$-bound		K$^+$-bound		K$^+$ unbound

The first transition from open to closed requires binding of H$^+$ (to a lysine residue in the M0 region of the N-terminus). The second transition requires K$^+$ binding in the pore region, with a conformation change in which residues in the so-called pore helix (I136, V140; Doyle *et al.*, 1999 & see below) play an important role (Schulte *et al.*, 2001), since appropriate mutation abolishes K$^+$ dependent gating. Schulte *et al.* (2001) also compare their description with that of C-type inactivation in voltage gated K$^+$ channels (Lopez-Barneo *et al.*, 1993). Since Kir2.0 (and most other subfamilies of K$^+$ channels) do not show slow changes in current upon changes of $[K^+]_o$, it is argued that these channel are not K$^+$-gated (Doi *et al.*, 1996; Pearson *et al.*, 1999; Schulte *et al.*, 2001). Certainly the process is not identical, but K$^+$ regulated conformation changes appear to occur in both cases.

Choe *et al.* (1998, 1999) also argued for a 'locus of' channel 'gating' in the pore region of inward rectifiers. They measured unitary currents in Kir1.1b and in Kir2.1. Measurements were made over a voltage range where polyamine occupancy of the channel would be negligible. Kir1.1b has short openings (23ms at -60mV, 110mM-K^+_o). Openings are long (280ms at -60mV) in Kir2.1 and the kinetics is similar to that described in inward rectifiers of skeletal myotubes (Matsuda & Stanfield, 1989). In the cloned channels, domain swaps were used to show that both the distribution of closed times and their duration were determined by the structure of M2 and of the extracellular loop that also includes the P-region. Paradoxically, since the opposite might be expected, introduction of the P-region of Kir2.1 (residues 132-148) into Kir1.1b *reduced* mean open time, while introduction of the P-region of Kir1.1b into Kir2.1 *increased* it. For the long open times found in Kir2.1, channels must possess the entire extracellular loop of Kir2.1, not just the P-region. An interaction with M2 seems likely however, since replacement of M2 in Kir2.1 by that from Kir1.1b reduces open time ten-fold (Choe *et al.*, 1999). Thus open time was reduced if *either* M2 *or* the extracellular loop of Kir1.1b replaced these domains in Kir2.1. Choe *et al.* (1998; 1999) argue for K^+ occupancy of a binding site in the pore resulting in conformation changes in the extracellular loop and perhaps in M2.

Two other groups have shown that other mutations in H5 alter channel kinetics at a single channel level (see Table 4 for sequences). Guo & Kubo (1998) found in Kir2.1 that mutation of Gln at position 140 to Glu (Q140E) reduced open time durations 10-fold, with the channels acquiring an additional fast component in the distribution of closed times. So *et al.* (2001) have also shown that mutation (of Tyr 145) in the pore region of Kir2.1 radically affects the duration of channel open times. These changes were made using tandem dimers in which one only of the pore regions was altered in structure. In particular, open time was reduced more than 10-fold by replacement of Tyr by Val (Y145V). So *et al* (2001) have therefore also argued for a conformation change in the pore region 'akin to C-type inactivation' of Kv channels occurring in Kir2.0.

Thus, first, the structure of the pore region (H5 or the selectivity filter) seems necessary for the precise matching of P_{open} to changes of K^+ concentration. An Arg residue (R148 in Kir2.1) is particularly important since its replacement by Tyr results in the loss of precise (V-E_K) dependence under conditions where $[K^+]_o$ is changed. The pore has a K^+ co-ordination site at this point and is sufficiently narrow here for K^+ to fit snugly. Secondly, this co-ordination site appears capable of regulating conformation changes that may involve the H5 region. These changes do not generate inward rectification however, and in Kir2.0 the changes have not yet been integrated into an overall picture of channel physiology. The K^+ dependent gating in Kir1.1, however, explains properties of K^+ secretion in renal cortical collecting duct (Giebisch, 1998).

3.6 Selectivity and Blockage of Inward Rectifier K+ Channels

3.6.1 Role of the Selectivity filter.

Many Kir channels are more selective, particularly between K^+ and Rb^+ (Adrian, 1964) than are Kv. Hagiwara & Takahashi (1974) were the first to measure channel selectivity in inward rectifiers of occytes by measuring the change in reversal potential, ΔE_{rev}, that follows replacement of K^+ with a test cation. This study gave permeabilities to various cations in the ratio Tl^+:K^+:Rb^+:NH_4^+ = 1.5:1.0:0.35:0.04 with undetectable permeance of Na^+ and Cs^+. In frog muscle, P_{Rb}/P_K was 0.66 (Standen & Stanfield, 1980). In Kv, P_{Rb}/P_K is often close to unity (Hille, 1992).

Permeability ratios have also been measured in cloned channels in expression systems. For example in Kir2.1, P_{Rb}/P_K was measured as 0.68 if Rb^+ is applied externally (Abrams *et al.*, 1996) but as 0.35 if the substitution is at the internal face of the ion channel (Reuveny *et al.*, 1996). This latter finding implies that there is also an inner selectivity filter, and this filter is primarily associated with the aspartate residue D172 in M2 (Reuveny *et al.*, 1996) that is also crucial to channel gating by polyamines.

The principal selectivity of Kir, as of voltage gated channels, is however conferred by the P- or H5-region of the channel, and this has a similar, though not identical structure to that found in Kv. Both classes of channels possess the T(s)xxT(s)xGY(f)G motif which appears virtually universal among tetrameric K^+ channels. The possible salt bridge between E138 and R148 in Kir2.1 (Yang *et al.*, 1997) has been discussed in the previous section.

An intersubunit disulphide bond between conserved Cys residues in the extracellular loops on either side of the P-region appears essential for the normal folding of Kir channels (Leyland *et al.*, 1999). Mutation of either Cys residue (C122 & C154 in Kir2.1) results in the lack of functional channel protein but in apparently normal cell trafficking. Activity, once the channel is folded, is not however affected by reducing agents that would break such a bond. The same finding was made in Kir2.3 (Bannister *et al.*, 1999). Cho *et al.* (2000) have confirmed the intrasubunit nature of the bond and its necessity for appropriate channel folding. Since the pair of Cys residues is completely conserved throughout the family, the requirement is almost certainly a property of all Kir channels.

The basic shape of the outer mouth of the channel seems similar to, but not identical with that of Kv. Although certain pore acting toxins do not bind to Kir, Lq2, a scorpion toxin from *Leiurus quinquestriatus*, will block Kir1.1a, but not Kir2.1 (Lu & MacKinnon, 1997). It also blocks K(Ca) and Kv, particularly Shaker (Lu & MacKinnon, 1997; Escobar *et al.*, 1993). Toxin binding is to the pore region, since mutations in the pore of Kir1.1a, including I142L,

Y144F, F146A and F148C (see Table 4 for the sequence of this region), and in the surrounding extracellular loops (especially mutations of E123 and N124) alter toxin affinity. Mutations of the toxin suggest that similar basic residues are important for binding to both Kv and Kir (Lu & MacKinnon, 1997). The toxin may therefore be expected to leave a footprint that is similar in each case. These findings suggest that structures solved for Kv channels will apply at least approximately to Kir. In turn, affinity for another toxin from *Leiurus quinquestriatus*, argiotoxin 1, is conserved in both eukaryotic and prokaryotic K^+ channels making the structure of the bacterial KcsA a good basis for consideration of structure of Kv and Kir (MacKinnon *et al.*, 1998). The basis of selectivity in Kir channels is likely to be common among known K^+ channels. In support of this assertion, the selectivity filter of KcsA may be engineered into Kir2.1, replacing the selectivity filter of this inward rectifier, without loss of ionic selectivity (Lu *et al.*, 2001).

One approach has been the use of scanning cysteine mutagenesis to attempt to identify residues whose side chains are pore lining in H5 (Dart *et al.*, 1998a, b; Kubo *et al.*, 1998). This approach had been used in Kv channels (Lü & Miller, 1995; Kürz *et al*, 1995; Pascual *et al.*, 1995), where residues were replaced by Cys and their accessibility to sulfhydryl reagents, such as Ag^+, Cd^{2+} or a range of methanesulphonate reagents, is examined. The approach in Kv suggested that Tyr in the K^+ channel motif is accessible and may line the pore. Similar results were found in Kir2.1 (Dart *et al.*, 1998b), a result consistent with the suggestion that K^+ permeates selectively through interaction with the molecular π-orbital of the aromatic ring of H5, much as predicted by Kumpf & Dougherty (1993). In summary, Dart *et al.* (1998b) found evidence suggesting that T141, T142, I143, Y145, F147 and C149 were exposed, whilst the side chains of T139, Q140, G144, and G146 were buried. The results were consistent with both Y145 and F147 in Kir2.1 contributing to selectivity through cation-π interactions. One difference between Kir and Kv was that T141 was found to be accessible to Ag^+ (Dart *et* al., 1998b), to chloramine T, and possibly to Cd^{2+} (Kubo *et al.*, 1998). The equivalent position in Kv1 and Kv2 was inaccessible to externally applied sulfhydryl reagents (Lü & Miller, 1995; Kürz *et al*, 1995), though that in Kv2 was accessible to internally applied reagents (Pascual *et al.*, 1995).

The hypothesis that cation-πinteractions form the basis for potassium ion selectivity seems likely to be incorrect. The X-ray crytallographic study of a bacterial K^+ channel KcsA (Doyle *et al.*, 1998) shows that Tyr of the K^+ channel motif points away from the pore as do several other residues whose side chains apparently become accessible when replaced by Cys. The result of Doyle *et al.* (1998) is that K^+ channel selectivity is conferred through backbone carbonyl oxygens as first postulated by Bezanilla & Armstrong (1972) and by Hille (1973). The structure has four such rings of oxygen atoms from the peptide bonds between the residues underlined in the KcsA sequence in Table 4. These rings form a tunnel through which K^+ must fit snugly, so that it can shed its water of hydration as it passes on its journey through this part of the channel. This tunnel is held at the correct diameter through interactions between the side chains of residues of the filter itself and those of a helical part of the P-region (the pore helix, shown italicised in the structures given in Table 4; see also Fig. 2C). These interactions are through van der Waal's forces and through a H-bond between Y78 of the selectivity filter and W68 (or possibly T72) of the pore helix of the adjacent subunit.

The structure of KcsA has four-fold symmetry - such symmetry was imposed on the modelling done to refine the structure found by X-ray crystallography. But four-fold symmetry is not a requirement for K^+ selectivity in Kir channels. Channels of the Kir3.0 subfamily are heteromers. Channels that allow the heart to respond with a reduction of rate to the action of acetylcholine (ACh) are made up as heteromers of Kir3.1 and 3.4 (Krapivinsky *et al.*, 1995a). As the sequences illustrated in Table 4 show, the structure of the P-region is not the same in these two channel subunits - both the selectivity filter and pore helix have a different sequence. Thus the channel is likely to show two-fold, but not four-fold symmetry. And the subunits appear to contribute differently to selectivity (Silverman *et al.*, 1998). Mutations were made in tandem dimers of Kir3.1 and Kir3.4, which will come together to make channels with subunits in the order 3.1,3.4,3.1,3.4. The wild type 1-4 constructs give a very good K^+ selectivity ratio, with a Na^+ current/ K^+ current ratio (I_{Na}/I_K) of 0.014. Mutations were made of Y in GYG of the K^+ channel signature sequence. Mutations in the Kir3.4 element of the tandem dimer had a greater effect on selectivity than did mutations in Kir3.1. Even the conservative mutation replacing Tyr with Phe had significant consequences for ionic selectivity if the replacement was in Kir3.4. Here, with Y152F in Kir3.4, $I_{Na}/I_K = 0.1$. Kir3.1 was tolerant of mutation of Tyr with many changes giving highly K^+ selective channels. Only the mutations to Asn, Arg, or Val gave a marked increase in relative Na^+ permeability. The Y→V mutant was measurably Ca^{2+} permeant, a finding in contrast to that in Kv, where the same mutation conserved high K^+ selectivity (Heginbotham *et al.*, 1994). Channels made of Kir3.4 alone gave small currents and channels were only moderately K^+ selective. Silverman *et al.* (1998) used chimeric channel proteins to test whether channels with four fold symmetry in the P-region (either all Kir3.1 or all Kir3.4 P-regions) were K^+ selective and found that they were not. Only channels constructed with a heteromeric pore were highly K^+ selective, so that the lack of four-fold symmetry is here a necessity for K^+ selectivity. It is also sufficient, since the constructs could possess identical non-pore sequences in each of the four subunits making up the channel. In the description of K^+ selectivity given by Doyle *et al.* (1998), residues making the selectivity filter must form specific side chain-side chain interactions with the pore helix in the *adjacent* subunit. The need for Kir3.1/3.4 ordering of pore regions is then consistent with this description if the selectivity filter of Kir 3.1 is matched to the pore helix Kir3.4 and *vice versa*.

In Kir channels the pore helix has a very different structure from that of KcsA or Kv channels. The couplet of Trp residues is absent from all Kir channels (Table 4) which have LeuPhe in its place. Tyr residues in GYG cannot then form a H-bond with Trp in the pore helix, though modelling suggests that a H-bond can be formed between Y145 and T139 in an adjacent subunit (So *et al.*, 2001). This H-bond is not essential for selectivity however since in Kir2.1 the mutant Y145F is not detectably different in either P_{Na}/P_K or P_{Rb}/P_K (So *et al.*, 2001). Thus van der Waal's forces may be sufficient to maintain the selectivity filter at its appropriate diameter.

Kir2.0, which are thought not to form heteromers within the subfamily (Tinker *et al.*, 1996), will possess pore regions with four-fold symmetry, but this symmetry is not essential for K^+ selectivity at the level of Tyr in GYG. So *et al.* (2001) made tandem dimers in which Tyr (Y145) was replaced by Ala, Met, Phe, Leu, or Val in one only of the two subunits. Surprisingly, none of these channels showed a difference in ionic selectivity from that of wild type channels (So *et al.*, 2001). In Kir2.1 at least, four-fold symmetry is not required in the pore region at the level of the ring of Tyr (Y145) residues.

Part of the early evidence against the hypothesis that cation-π interactions were responsible for selectivity came from the finding that replacement of G in GYG of Kv channels had a greater effect on selectivity than did replacement of Y (Heginbotham *et al.*, 1994). It is consistent with the structure of Doyle *et al.* (1998) that mutations of G remove K^+ selectivity: Gly residues are required for rotational flexibility around the peptide bond to permit the backbone carbonyl oxygens to point the same way – into the pore. Murine Kir3.2 has a natural mutation G156S in G^{156}YG (Patil *et al.*, 1995). The mutant is the *weaver* mouse, owing to its movement disorders, including cerebellar ataxia. The mutation gives rise to currents that are non-selective among monovalent cations whether the channel is expressed in homomeric or heteromeric form (Kofuji *et al.*, 1996b; Navarro *et al.*, 1996; Slesinger *et al.*, 1996; Hou et al., 1999). For example in Kir3.1/*wv*Kir3.2 heteromers, P_{Na}/P_K was 0.74 and P_{Cs}/P_K was 0.93 (Navarro *et al.*, 1996). Further, channels lost their sensitivity to Gβγ (Navarro *et al.*, 1996) so that the currents became constitutively or basally active (Kofuji *et al.*, 1996b; Slesinger *et al.*, 1996). This basal activity caused cell death in CHO cell and oocyte expression systems (Navarro *et al.*, 1996) and is the probable cause of the degeneration that occurs of cerebellar granule cells. Kir3.2 is expressed at high level in granule cells during cerebellar development in the first post-natal days (Slesinger *et al.*, 1996). Cerebellar granule cells in primary culture appear to develop normally if Na^+ influx through the mutant Kir channels is blocked by the local anaesthetic QX314 or the NMDA receptor blocker MK-801 (Kofuji *et al.*, 1996b), but neurite outgrowth appears to be blocked by the background Na^+ current. A similar G→S mutation in Kir2.1 is one of the identified causes of Andersen's syndrome in humans (Plaster *et al.*, 2001).

Thus it is believed that ionic selectivity depends on structures similar to those described in KcsA (Doyle *et al.*, 1998). The basis of selectivity is a tunnel of backbone carbonyl oxygen atoms held in place by side chain – side chain interactions between residues in the selectivity filter and in the pore helix of the adjacent channel subunit.

3.6.2 Blockage by Foreign Cations Is Influenced by Several Sites Through the Pore

Experiments using extracellular blocking cations were important initially in establishing Kirs as channels rather than carriers. Blocking cations were then used to separate Kir currents from other currents in native tissues, and finally to help establish newly cloned channels as members of the Kir family. There has therefore been some effort to identify those residues that confer high affinity blockage by Cs^+, Ba^{2+}, etc. Although the expression used to fit voltage-dependent blockage implies that a blocking cation will get only so far through the channel and then get stuck, any residue may affect blocking kinetics and/or affinity. Thus, several sites appear to be involved, some of them at the external mouth of the channel (Alagem *et al.*, 2001), some in the P-region (Sabirov *et al.*, 1997b; Zhou *et al.*, 1996; Thompson *et al.*, 2000; Alagem *et al.*, 2001) and some at least in the inner, wider part of the pore (Abrams *et al.*, 1996; Thompson *et al.*, 2000).

Cs^+, Rb^+ (Abrams *et al.*, 1996), and Ba^{2+} (Shieh *et al.*, 1998) block Kir2.1 channels in much the same way that they block native strong inward rectifiers, with similar competition between permeant and blocking cations. Intracellular spermine interferes with this competition, slowing recovery from Ba^{2+} blockage at positive membrane potentials in excised, inside-out membrane patches (Shieh *et al.*, 1998).

Residues at the outer pore mouth may affect kinetics of blockage. First, an Arg residue at the mouth of the pore (R148 in Kir2.1) affects blockage by extracellular cations. Sabirov *et al.* (1997b) made the mutation R148H. They were unable to express this mutant by itself, but obtained currents different from those associated with wild type channels if both wild type and mutant mRNA were injected into *Xenopus* oocytes. Though the channel population is expected to be a mixture of WT homomers and WT/R148 heteromers (of differing stoichiometries), useful information came from the study. The mutation resulted in an acceleration of on-rate for both Cs^+ and Ba^{2+}, but with little change in affinity or voltage dependence of steady state blockage. Similarly, in Kir2.1, a glutamate residue in the extracellular loop between M1 and H5 also affects blocking kinetics by Ba^{2+}. Mutation of this residue to Asn (E125N) slows the development of blockage by Ba^{2+}, without affecting affinity (Alagem *et al.*, 2001). Acidification, to protonate E125 has a similar effect on rate of development of blockage. In Kir7.1, mutation to Arg of the Met residue in the equivalent position to R148 of Kir2.1 results in an increase in blocking affinity for Ba^{2+} (Krapivinsky *et al*, 1998b; Döring *et al.*, 1998).

Within the selectivity filter, differences in blocking affinity between Kir1.1a and Kir2.1 have been explained by residue V140 in Kir1.1a and T141 in Kir2.1 – these are in the equivalent position (Table 4). Mutation V140T enhances Ba^{2+} blockage in Kir1.1a (Zhou *et al.*, 1996); mutation T141V in Kir2.1 reduces blocking affinity by Cs^{+} (Thompson *et al.*, 2000) and by Ba^{2+} (Alagem *et al.*, 2001) in Kir2.1. This residue lies immediately below the backbone carbonyl oxygen tunnel that confers selectivity (see Doyle *et al.*, 1998). It is then likely that blocking cations traverse this tunnel.

Perhaps surprisingly, however, residues in M2 also appear to affect affinity. In particular the mutant S165L in Kir2.1 (Table 2) was found to be very insensitive to blockage either by Cs^{+} or Rb^{+} (Thompson *et al.*, 2000). Rb^{+} currents, which are normally small, owing to Rb^{+} blockage, became larger than K^{+} currents. Thompson *et al.* (2000) considered whether the mutation was acting through a change in the structure of the selectivity filter. The hypothesis was that S165 formed a H-bond with T141 to help stabilise the selectivity filter at the appropriate diameter. Consistent with this hypothesis, mutations of T141 to polar residues had little effect, but all mutations of T141 to non-polar residues reduced blockage. Thus it is likely that T141 makes side chain – side chain interactions to stabilise the selectivity filter. However, the mutation S165A and other mutations that would disrupt H-bond formation had little effect on blockage. Thus, any interactions of T141 are unlikely to be with S165. Indeed, Minor *et al.* (1999) have evidence that S165 lines the wide inner pore of Kir2.1 (see Table 2).

A double mutation in M2 of Kir2.1 (S165L/D172N) produces channels which carry larger Rb^{+} and large Cs^{+} currents at negative voltages (Thompson *et al.*, 2000) indicating that blocking cations probably traverse the selectivity filter to reach their blocking site. These mutations do not alter permeability ratios, measured from the change in reversal potential upon replacing K^{+} by the blocking cation. The selectivity filter (the H5 or P region) determines these ratios. Rather, these mutations determine whether ions, once they have entered the channel, are able to move through to the intracellular milieu. And, if ions are able to move through, residues in M2 determine the rate at which such movement occurs.

3.7 Modulation of channels of the Kir2.0 subfamily

Unlike channels of the Kir3.0 subfamily, Kir2.0 channels do not require activation by G-proteins (see below). Perhaps for this reason, their modulation and their role as targets of cell signalling pathways activated by neurotransmitters or hormones has, with the exception of Kir2.3, been somewhat neglected. However, Kir2.0 channels are bound in signalling complexes with scaffolding proteins, and constitutively active inward rectifiers are down-regulated in a number of physiological situations.

Like certain voltage-gated potassium channels (Kim, E. *et al.*, 1995), Kir2.1, 2.2, and 2.3, but not 2.4, possess a motif at the C-terminus denoting their ability to be bound to the cytoskeletal PDZ domain or MAGUK (membrane associated guanylate kinase) family of proteins. The motif is also present in Kir3.2c, a splice variant of Kir3.2 (see Table 1), and in channels of the Kir4 subfamily. The acronym PDZ denotes a family of proteins that includes the post-synaptic density protein PSD-95, the Drosophila septate junction protein, discs-large, and the epithelial tight junction protein, zona occludens-1. Proteins that have the sequence T/S-x(D/E/A)-V/I at the C-terminus bind to PDZ domain or domains in such proteins (e.g. Cohen *et al.*, 1996b; for review, see Pawson & Scott, 1997). This C-terminus sequence is also termed the C-TAG (C-terminal association with GLGF, the PDZ domain sequence) motif. In Kir2.1 and 2.2, the C-terminal sequence of the channel protein is RRESEI; in Kir2.3, it is RRESAI. In both cases the C-TAG motif overlaps a PKA phosphorylation consensus sequence (RRxS) and association of Kir2.3 with PDZ-domain proteins is regulated by phosphorylation and dephosphorylation (Cohen *et al.*, 1996b). The binding to PDZ-domain proteins is known to underlie clustering of Kir2.2 in transverse tubular systems of ventricular myocytes (Leonoudakis *et al.*, 2001) and of Kir4.1 channels in retinal Müller cells (Horio *et al.*, 1997; Ishii *et al.*, 1997).

Kir2.3 was the first to be shown by immuno-coprecipitation to be associated with PSD-95 in forebrain, including hippocampus (Cohen *et al.*, 1996b). This association failed if Glu, Asp or Ala replaced Ser at the C-terminus by site-directed mutagenesis. PKA-dependent phosphorylation detached the channel from PSD-95. Forskolin reduced the amount of Kir2.3 immuno-coprecipitated with PSD-95 and phosphatase action restored association (Cohen et al. 1996b). Kir2.1 has also been shown to associate with PSD-95 (Nehring *et al.*, 2000). Nehring *et al.* (2000) used the yeast two hybrid method to assess the binding of PSD-95 by the C-termini of a number of Kir channels, finding that Kir2.1 and 2.3 bound this protein, but Kir3.0 channels did not. They also showed immuno-coprecipitation of both Kir2.1 and 2.3 with PSD-95. Co-expression of Kir2.3 with PSD-95 in COS-7 cells resulted in channel clustering, but co-expression in HEK-293 cells resulted in a reduction in Kir2.3 current. This reduction was shown to be associated with a reduction in unitary conductance (high symmetrical [K^+]) from 14 to 8pS. An internal PDZ recognition motif (ETGI) seemed to be important in this effect (Nehring *et al.*, 2000).

Outside the central nervous system, other MAGUK proteins are involved in anchoring Kir2.0. Leonoudakis *et al.* (2001) used GST-fusion proteins with the C-terminal regions of Kir2.1, 2.2, and 2.3 to find novel interacting proteins. SAP97 bound all three channel proteins, with Kir2.2 binding the second of three PDZ domains of SAP97. The association was inhibited by protein kinase A phosphorylation of Kir2.2. It was abolished if the C-terminal triplet of residues (SEI) was cut from the Kir2.2 fusion protein. In both cerebellar glial cells (Bergmann glia in the molecular layer, astrocytes in the granule cell layer) and in the T-system of ventricular myocytes, Kir2.2 co-localised with SAP-97. Kir2.1 was also reported to have a t-tubular location (Leonoudakis *et al.*, 2001). Kir4.1 and 4.2 proteins are also associated in the central nervous system with CIPP (channel-interacting PDZ domain protein) which also binds glutamate receptors of the NMDA class (Kurschner *et al.*, 1998).

PKA action has been described as necessary for (Fakler *et al.*, 1994b), as enhancing (Dart & Leyland, 2001) and as inhibiting Kir2.1 function (Wischmeyer & Karschin, 1996; Jones, 1997). In cardiac ventricle, channels carrying I_{K1} (principally Kir2.1; Zaritsky *et al.*, 2001) are inhibited by the β-adrenergic agonist isoproterenol (Koumi *et al.*, 1995a,b). This effect of isoproterenol is mimicked by forskolin and dibutyryl cAMP, and is abolished by PKI(5-24), a peptide specific inhibitor of PKA, suggesting that the reduction of Kir current originates from a PKA-dependent phosphorylation of Kir. Similarly, Wischmeyer & Karshin (1996) showed that $5HT_{1A}$ receptors could cause inhibition of Kir2.1 channels in COS-7 cells in which receptor and channel had been co-expressed. Although $5HT_{1A}$ receptors are normally thought to link negatively to adenylyl cyclase (Watson & Arkinstall, 1994; see however Lucas *et al.*, 1993), there was evidence that in this case inhibition was due to stimulation of PKA. Inhibition was mimicked by addition of cAMP to the cell interior or by the catalytic subunit of PKA. Removing the Ser residue from the PKA consensus sequence (mutant S425N) in Kir2.1 prevented inhibition. The mutant was reported to express at levels comparable to those found with wild type in spite of its lack of a C-TAG motif. Inhibition by PKA in this instance therefore seems unlikely to be associated with the breaking of a link to PSD-95 or other PDZ domain protein. Previously, it had been argued that PKA phosphorylation was required for high level of channel function in a *Xenopus* oocyte expression system (Fakler *et al.*, 1994b), with PKA phosphorylation reversing channel rundown (Ruppersberg & Fakler, 1996).

This complexity in the action of PKA is puzzling, but may depend on how Kir2.1 relates in the cell membrane to other cell signalling proteins in the particular expression system used. Dart & Leyland (2001) have recently shown that Kir2.1 may be associated in cell membrane with the A kinase anchoring protein AKAP79 (also known as AKAP5). This protein targets PKA action to particular cellular locations (*e.g.* Dell'Acqua & Scott, 1997). In HEK 293 cells, Kir2.1 currents were enhanced by application of cAMP if Kir2.1 was co-expressed with AKAP79. The anchoring protein significantly increased the response to PKA. But this effect was not only dependent on AKAP79, it also required phosphatase action to be blocked. PKA phosphorylation acted to shift the channel open state probability to more positive voltages, suggestive of an interaction between phsophorylation and gating by polyamines. HA-tagged Kir2.1 and AKAP79 were immuno-coprecipitated with antibodies directed against the Kir2.1 HA epitope tag. Kir2.1 and AKAP79 co-localised in cell surface membrane. GST-fusion proteins were used to show that AKAP79 could bind to both the intracellular N- and C-terminal domains. Binding in the C-terminus did not involve the PDZ binding motif (Dart & Leyland, 2001).

Modulation of Kir2.3 has been extensively studied, as these channels are down-regulated by a number of factors. These factors include: protein kinase C dependent phosphorylation (Henry *et al.*, 1996; Zhu *et al.*, 1999b); intracellular Mg^{2+} (Chuang *et al.*, 1997); interaction with $G\beta\gamma$ (Cohen et al., 1996b); ATP (Collins *et al.*, 1996); changes of intra- and/or extracellular pH (Zhu *et al.*, 1999a); and generation of reactive oxygen species (Duprat *et al.*, 1995a). Kir2.3 appears to be directly stimulated by arachidonic acid (Liu *et al.*, 2001). Apart, perhaps, from PKC-dependent phosphorylation, most of these factors do not affect Kir2.1 activity.

Strong inward rectifiers possess several sites for PKC phosphorylation. Phosphorylation by PKC is reported to inhibit Kir2.1 (Fakler *et al.*, 1994b). Co-expression of m1 muscarinic receptors with Kir2.1 channels in a mammalian cell line leads to down regulation of these K^+ channels by carbachol. This action is blocked by atropine and mimicked by the phorbol ester, phorbol-12,13-dibutyrate (Jones, 1996; see also Jones 1997).

It is well established that Kir2.3 is inhibited by PKC-dependent phosphorylation. Using oocyte expression, both Henry *et al.* (1996) and Zhu *et al.* (1999b) found that whole cell Kir2.3 currents, but *not* those of Kir2.1, are inhibited by phorbol 12-myristate14-acetate. This sensitivity was lost upon replacing the N-terminus with that of Kir2.1 (Zhu *et al.*, 1999b). Serine residues in consensus sequences in the N-terminus of Kir2.3 are at positions 5, 36, 39 and 53. Only mutation of Ser53 abolished PKC action on Kir2.3. Introduction of a PKC site in an equivalent position in Kir2.1 conferred phorbol sensitivity on this channel when expressed in oocytes (Zhu *et al.*, 1999b). As with PKA (see above), there is again controversy over the possible inhibitory effect of PKC on Kir2.1

Kir2.1 is the target of phosphorylation by tyrosine kinases, which reduce channel activity (Wischmeyer *et al.*, 1998). This effect is the likely route through which NGF and EGF inhibit Kir2.1 after expression in COS-7 cells or in oocytes (Wischmeyer *et al.*, 1998). Perorthovanadate, a tyrosine phosphatase inhibitor, reduced Kir2.1 currents; this inhibition was abolished if tyrosine kinase activity was blocked by genistein. The mutant Y242F was not affected in this way, indicating that this is the site on the channel C-terminus that is phosphorylated (Wischmeyer *et al.*, 1998). It is however not certain that phosphorylation of Y242 simply inhibits Kir2.1 channels. In oocytes, the action of tyrosine kinase induces very substantial clathrin mediated endocytosis, and thereby removes channels from the surface membrane (Tong *et al.*, 2001).

Chuang *et al.* (1997) have proposed m1 receptor modulation of Kir2.3 (but not of 2.1) occurs through changes in intracellular free Mg^{2+} in *Xenopus* oocytes, rather than through PKC. Kir2.3 is inhibited by carbachol if m1 receptor and channel are co-expressed. This inhibition is not due to PKC-dependent phosphorylation. Neither a PKC peptide inhibitor (PKC 19-36) nor direct application of PKC mimics the effect of carbachol. Chimeric (Kir2.1/Kir2.3) channels show that elements of both the N and the C-termini of Kir2.3 are required for inhibition, but inhibition occurs in chimeras that lack any of the PKC consensus sequences of Kir2.3. Nor do other second messengers (including diacylglycerol, IP_3, or Ca^{2+}) mimic the effect. PIP_2 depletion was not considered, but would have been likely to affect both Kir2.1 and 2.3. Inhibition still occurs in the Kir2.1 mutant D172N/E224G that has low affinity for pore blockage by polyamine or Mg^{2+}. Intracellular Mg^{2+} therefore inhibits channels after application of carbachol has raised its concentration, and does so through a route other than pore blockage. Measured dose-response curves for this effect of Mg^{2+} permit estimation of the change in Mg^{2+} concentration caused by carbachol. The estimates (a rise of ~200μM) agree among three channel types (wild type Kir2.3, mutant Kir2.1 and one chimera) that have different affinities for Mg^{2+} (Chuang *et al.*, 1997). Single channel recording suggests that Mg^{2+} puts the channels into a long-lived inactivated state. Thus a change of free intracellular Mg^{2+} may provide a mechanism for channel inhibition.

Interaction with Gβγ also inhibits Kir2.3, but not Kir2.1 (Cohen *et al.*, 1996b). Channel activity was suppressed by co-expression of Kir2.3 with $G\beta_1$ and $G\gamma_2$. Some reduction in activity was also found with Gαi and Gαq, but not with Gαs or Gαt. This element of channel gating works through the N-terminus, unlike the situation with Kir3.0 (see next section; Cohen *et al.*, 1996b).

Kir2.3 channels also appear to be regulated by ATP (Collins *et al.*, 1996), a property most notably associated with Kir6.0/SUR (K_{ATP}-sulphonylurea receptor complexes). In excised, inside-out membrane patches, Kir2.3 channels are inhibited by 5mM-ATP, though the inhibition takes several s to come to completion. As with K_{ATP}, inhibition is independent of the presence of Mg^{2+} and is mimicked by non-hydrolysable analogues of ATP (Collins *et al.*, 1996). And, as with K_{ATP}, the inhibitory effect is reversed by the presence of Mg-ADP. Unlike K_{ATP}, ATP affinity is low, with inhibition being generated by mM, rather than μM concentrations of the nucleotide. Further, unlike K_{ATP}, where association with SUR is required for the stimulatory effect of Mg-ADP, there is no evidence of association of Kir2.3 with SUR or with any other ABC-protein.

Kir2.3 also appears to be exquisitely sensitive to changes of extra- and intracellular pH over the physiological range (Coulter *et al.*, 1995). Channels are shut by acidification, giving a possible mechanism for an excitatory response of neurones to hypercapnia (Zhu *et al.*, 1999a; Zhu *et al.*, 2000). Changes of pH_o resulted in a reduction by up to 50% of unitary current, the effect occurring with a pK_a of 6.7 and a Hill coefficient of 3 – 4. The effect was independent of voltage (Zhu *et al.*, 1999a). Changes of pH_i had effects both on channel P_{open} and on unitary current – two separable actions. The principal effect was a reduction in P_{open}, which occurred with a pK_a of 6.8 – the dependence was steep, with a Hill coefficient of 5. The reduction in unitary current was again incomplete (~40%) at acid pH_I and had a pK_a of 6.7. Reductions in unitary current are likely to involve pore blockage at inner and outer pore mouths (Zhu *et al.*, 1999a). The reduction in P_{open} was later shown to involve a gating process (Qu *et al.*, 2000). Kir2.1 is affected by changes of pH_o and pH_i only at very low values of pH (Sabirov *et al.*, 1997a). Kir2.4 is sensitive to changes of pH_o within the physiological range (Hughes *et al.*, 2000).

Qu *et al* (1999) used domain swaps of N- and C-termini between Kir2.1 and Kir2.3 to search for residues that regulated the effect of intracellular protons on P_{open}. They found that N-terminal substitution eliminated pH sensitivity. Several residues in the region of the N-terminus close to M1 seemed to be the most important. Introduction of Kir2.1 residues 77 – 86 (the last 10 residues before the start of M1 and therefore part of M0) into the equivalent position in Kir 2.3 (positions 51-60) had major effects on pH sensitivity (Qu *et al.*, 1999). Qu *et al* (1999) believed Thr53 (which confers the response to phosphorylation by PKC) to be a residue of major importance, since its mutation to Ile had substantial effect. The double mutation T53I/Y57W further reduced pH-sensitivity.

In addition, Qu *et al.* (2000) found that a cumulative effect of replacement of several His residues in the C-terminus converted the sensitivity to changes in pH_i from that characteristic of Kir2.3 into that of Kir2.1. Gating was believed to involve a pH-dependent interaction between N- and C-termini. A region of the C-terminal from residues 196 – 230 was of importance – similar to the region of the C-terminus that Lu *et al.*, (1999b) described as contributing to the cytoplasmic vestibule at the inner pore mouth. Within this region a motif `PYMQ` was critical (Qu *et al.*, 2000), since mutation of these residues abolished pH sensitivity. In addition purified N- and C-termini bound each other *in vitro* and this interaction was stronger at pH 6.6 than at 7.4. Two regions are important, then, for the change in P_{open} upon lowering pH_i: the residues immediately before M1 (and particularly T53 and Y57; Qu *et al.*, 1999) and the motif `PYM(T)Q` (residues 226-230; Qu *et al.*, 2000). These regions may be responsible for an interaction between N- and C-termini gating the channel shut. This channel closure is more likely to occur if multiple titratable His residues in the C-terminus are protonated (Qu *et al.*, 2000).

Kir2.3 is inhibited by the release of reactive oxygen species through the photoactivation of rose bengal, a generator of singlet oxygen molecules (Duprat *et al.*, 1995a). Kir2.1 and Kir1.1a were not inhibited, but the response was shared with a number of Kv channels, (Kv1.3, 1.4, 1.5 and 3.4). Kir2.3 was not affected by agents generating primarily superoxide or hydroxyl radicals (Duprat *et al.*, 1995a).

Kir 2.3, but not 2.1, 2.2 or 2.4, is strongly stimulated by arachidonic acid (Liu *et al.*, 2001) as are certain other K^+ channels (Meves, 1994). This effect occurs through extracellular or intracellular application to channels expressed in CHO cells, but arachidonic acid (AA) is believed to act at an extracellular site since intracellular buffering of AA does not reduce the action. Currents were increased up to 2.5 fold by AA, which acted with an EC_{50} of about 450nM. Inhibitors of AA hydrolysis do not alter stimulation, and certain non-metabolised fatty acids mimic AA. The effect was shown not to be through PKC activation or the action of free oxygen radicals (Liu *et al.*, 2001) – in any event these agents have inhibitory effects on Kir2.3 (Henry *et al.*, 1996; Zhu *et al.*, 1999; Duprat *et al.*, 1995). Chimeric constructs of Kir2.1 and 2.3 were used to investigate those parts of the channel that conferred the response to AA. Chimeras containing only the N-terminus & M1 or M2 & the C-terminus of Kir2.3 fail to respond. Both M1 and M2 (and the N- and C-termini) of Kir2.3 seemed to be required for the channel to be stimulated by AA. Since Kir2.3 is strongly expressed in heart, this response may explain the reduction in electrical excitability of rat neonatal cardiac myocytes caused by AA (Kang *et al.*, 1995).

Thus, several pathways lead to the modulation of strong inward rectifiers, though Kir2.3 appears to be much the most responsive to intracellular signalling pathways. Descriptions of the response of Kir2.1 to phosphorylation by protein kinase A or C vary and the response may depend on how Kir2.1 is integrated into signalling complexes in the expression system used.

Kir2.0 channels are strong inward rectifiers that are constitutively active in cell membranes. Their open state probability may, however, be modulated by a number of cell signalling processes. Kir3.0 channels show little constitutive activity. They are also strong inward rectifiers, but they require interaction with G-proteins for high activity. These channels are discussed in the next section.

4 G-Protein Coupled Inward Rectifier K^+ Channels Belong to the Kir3.0 Subfamily

The activity of G protein-coupled inward rectifier K^+ (Kir3.0, GIRK) channels is primarily regulated by hormones and neurotransmitters. Inhibitory transmitters (such as somatostatin) activate the receptor, resulting in the dissociation of the G protein into the α- (Gα) and the βγ-subunits (Gβγ); Gβγ, in turn, increases activity of Kir3.0 channels. In this way, the channels produce a hyperpolarization in response to the signals coming from the transmitters. Thus, Kir3.0 channels regulate cellular excitability in the face of changing chemical environments.

In 1993 Kubo *et al.* (1993b) and Dascal *et al.* (1993) succeeded in cloning an ACh-activated Kir channel (K_{ACh}) from heart cDNA libraries. Kubo *et al.* (1993b) achieved this by the PCR method using degenerate primers based on the DNA sequence of Kir2.1 (Kubo *et al.*, 1993a). Dascal *et al.* (1993) succeeded by using the technique of expression cloning. The cloned DNA is called GIRK1 (Kir3.1). Injection of cRNA derived from this clone into *Xenopus* oocytes results in the expression of channels that behave like the native K_{ACh} channel. Kir3.1 is expressed in the atria and in other tissues including the brain (Kubo *et al.*,1993b; Dascal *et al.*, 1993).

Two additional cDNAs belonging to the Kir3.0 subfamily have been cloned (Lesage *et al.*, 1994); they are Kir3.2 (GIRK2) and Kir3.3 (GIRK3). These clones were obtained from a brain cDNA library by low stringency hybridization using Kir3.1 as the probe. These genes are expressed mainly in the CNS (Lesage *et al.*, 1994; 1995). Krapivinsky *et al.* (1995a) determined that Kir3.4 (CIR), together with Kir3.1, makes up the atrial Kir3.0 channel (K_{ACh} channel).

4.1 Atrial K_{ACh}

Stimulation of the vagus nerve inhibits atrial myocytes. Burgen and Terroux (1953) and Trautwein and Dudel (1958) showed that the inhibition of atrial myocytes is due to an increase in K^+ conductance caused by ACh released from the nerve terminals. Three decades later, Sakmann *et al.* (1983) discovered that there are two types of inwardly rectifying K^+ currents in heart pacemaker cells: the ACh-activated K^+ current (K_{ACh}) responsible for the inhibition by the vagus nerve, and the classical K^+ channel (which corresponds to Kir2.0). At the single channel level, the K_{ACh} channel has a much shorter open time (1-2 ms) than does the classical Kir channel (10-100 ms). This report by Sakmann *et al.* (1983) was the first description of Kir3.0 channels, though no reference was at that time given for the involvement of G proteins. Subsequently, Soejima and Noma (1984), using the technique of perfusing the solutions inside the patch electrode, presented strong evidence that the effect of ACh on K_{ACh} channels is mediated locally, in a membrane-delimited fashion, and not by diffusible second messengers.

Over the next few years, the mechanism of K_{ACh} activation was clarified owing to the publication of several important papers. Two reports, simultaneously published, analyzed the ACh-induced activation of K_{ACh}, and established that, unlike the case of the nicotinic ACh receptor, the muscarinic receptor and the K_{ACh} channel are two separate entities, with the G protein playing the role of intermediary (Pfaffinger *et al.*, 1985; Breitwieser and Szabo, 1985). Pfaffinger *et al.* (1985) showed that ACh induces K_{ACh} current only in the presence of intracellular GTP. Pertussis toxin abolishes the ACh-induced activation of K_{ACh}. Breitwieser and Szabo (1985) indicated that intracellular application of the non-hydrolyzable GTP analogue GppNHp gives an ACh-induced persistent activation of K_{ACh} (see also Endoh *et al.*, 1985; and Sorota *et al.*, 1985).

Kurachi *et al.* (1986) were the first to report single channel recordings using excised membrane patches from atria cells to study involvement of G proteins. In the presence of an agonist (ACh or adenosine) on the external side of the cell, the K_{ACh} channel is activated only when the cytoplasmic side contains GTP. This activation is abolished by pertussis toxin. The non-hydrolyzable GTP analogue GTPγS causes persistent activation of the K^+ channel. Next, Yatani *et al.* (1987) demonstrated that application of a purified G protein to excised patches of atrial cell membrane induces K_{ACh} activity. These experiments provide further evidence for the idea that a G protein plays an essential role in the K_{ACh} activation, and that the whole process of signalling occurs without the participation of diffusible messengers.

4.2 Gβγ Activates Kir3.0

Once the role of a G protein as a transducer for K_{ACh} activation became clear, the next step was to determine which subunit (α or βγ) of the G protein is responsible for activation. In 1987, Logothetis *et al.* provided evidence that the βγ subunit activates the K_{ACh} channels. This report was the first to indicate that Gβγ can be an agent acting on any effector. Their experiments were performed by applying purified G protein subunits directly onto the inside-out membrane patches of atrial cells. Their main results were these: K_{ACh} channels are activated by Gβγ (23 nM) but are not activated by either Gαi (Gα-GDP or Gα-GTPγS at concentrations up to 64 nM) or the trimer Gαβγ. At the time of the experiments, for any signalling pathways involving G proteins, the Gα subunit (not Gβγ) was believed to be the agent acting on the next effector. Against this background, the conclusion of Logothetis *et al.* (1987) was viewed with cepticism. Adding to the controversy, Codina *et al.* (1987) reported a few months later that activated Gα (Gα-GTPγS) at a very low concentration (5 pM) induces K_{ACh} activation, while Gβγ is ineffective at 2-4 nM. The conclusion of Codina *et al.* (1987) was that Gα is the primary agent activating the K_{ACh}.

It is now established that the Gβγ subunit activates Kir3.0 channels. The data supporting this conclusion are as follows. (1) Gβγ at nanomolar concentrations activates K_{ACh} very consistently (97 % of the patches; Logothetis *et al.*, 1988; 99%, Ito *et al.*, 1992). Activated Gαi (i.e., Gαi-GTPγS) does induce channel activation (at picomolar range); however, this activation is less consistent (obtained in only 20-50 % of the patches, Logothetis *et al.*, 1988). Even at a saturated dose, the degree of activation is only about 20 % of that induced by Gβγ (Ito *et al.*, 1992). (2) Despite the fact that Gαi activates K_{ACh} poorly, Gαi does activate the K_{ATP} channel of ventricular myocytes vigorously and consistently (Ito *et al.*, 1992); this result verifies that the purified Gαi protein used is functional (Ito *et al.*, 1992). (3) Application of GTPγS to the membrane activates K_{ACh} ; this activation must be induced by an endogenous G protein. This response is deactivated by Gαi-GDP, indicating that the exogenously applied Gαi sequesters endogenous Gβγ, inactivating its action. Thus, endogenous Gβγ (not Gα) must be the activator of K_{ACh} (Logothetis *et al.*, 1988). (4) Application of recombinant G protein subunits, made by baculovirus vectors in an insect (Sf9) cell line, produces essentially the same results as those obtained with the purified native subunits (Wickman *et al.*, 1994). Recombinant Gαi, however, does not activate K_{ACh} even at a high concentration (100 nM). (5) In contrast to experiments where purified G proteins were applied to excised patches, Reuveny *et al.* (1994) expressed G protein subunits and Kir3.0 channels in *Xenopus* oocytes by injecting cRNAs. When cRNAs for Kir3.1, $G\beta_1$, and $G\gamma_2$ (but not Gα) are injected into oocytes, spontaneous channel activity originating from Kir3.1 is observed with on-cell patch and single cell recording. This activity does not decrease even when the patches are excised and the cytoplasmic side is perfused with a GTP-free solution. Thus the activity is not due to the dissociation of the G protein into the Gα-subunit and the Gβγ-subunit, but originates from the over-expressed $G\beta_1\gamma_2$. This spontaneous Kir3.1 channel activity is however drastically decreased by application of a recombinant Gαi, indicating that the activity stops when $G\beta_1\gamma_2$ is sequestered. When cRNA for $G\alpha i_2$ is injected, in addition to cRNAs for Kir3.1, $G\beta_1$, and $G\gamma_2$, spontaneous channel activity is observed in the on-cell mode. This Kir3.1 activity must come from GTP-dependent dissociation of Gα from Gβγ, since activity ceases upon patch excision into a GTP-free solution. Subsequent application of an activated recombinant Gαi (50 nM) does not further activate channels. These data convincingly show that Gβγ, not Gα, is the agent that activates Kir3.0.

One question remains unanswered: what is the significance of K_{ACh} activation by a low concentration (picomolar) of purified Gαi (Codina *et al.*, 1987)? Activation of the K_{ACh} channel by picomolar Gαi, confirmed by Logothetis *et al.* (1988) and by Ito *et al.* (1992), cannot be explained by contamination with GTPγS (Wickman *et al.*, 1994). It remains a possibility that purified native Gαi is different in a subtle way from recombinant Gαi, and that during the natural *in vivo* activation of Kir3.0 channels, Gαi may play more important roles than are revealed by the experiments described above.

4.3 Kir3.0: Heterotetramers

Initially, it was presumed that a functional Kir3.0 channel is, by analogy with Kir2.0 channels, composed of identical subunits (homomultimers). Soon it was realized that Kir3.0 channels are heteromultimers.

Krapivinsky *et al* (1995a) found a protein in atrial tissues that co-immunoprecipitates with Kir3.1. Cloned cDNA encoding this protein is called CIR (Kir3.4, GIRK4). Coexpression of both Kir3.1 and Kir3.4 in *Xenopus* oocytes gives a robust Kir channel activity, while expression of Kir3.1 alone induces less activity, and expression of Kir3.4 alone fails to produce Kir3.0-like activity (Krapivinsky *et al.*, 1995a). Their conclusion is that K_{ACh} is a heteromultimer of two different subunits, Kir3.1 and Kir3.4. Later, a CIR homologue (XIR; Kir3.5) was found to be endogenously expressed in *Xenopus* oocytes (Hedin, 1996), and this is the reason why injecting Kir3.1 cRNA alone into oocytes produces a modest Kir3 channel activity. Essentially the same results were obtained by Duprat *et al.* (1995b).

When Kir3.1 alone or Kir3.2 alone is expressed in *Xenopus* oocytes, only a small current can be activated, and Kir3.3 alone yields no current. However, if either Kir3.2 or Kir3.3 is expressed together with Kir3.1, G protein-mediated currents are increased 7 to 40 fold over those found with Kir3.1 expression alone (Kofuji *et al*, 1995). Interestingly, expression of Kir3.3 suppresses the current generated by Kir3.2 alone (Kofuji *et al*, 1995). Velimirovic *et al.* (1996) used both *Xenopus* oocytes and a mammalian cell line (CHO cells) for expressing members of the Kir3.0. A combination of Kir3.1/Kir3.2 or Kir3.1/Kir3.4 produces currents up to 10-fold larger than do any of the subunits expressed alone (see also Duprat *et al.*, 1995b; Ferrer *et al.*, 1995; and Spauschus *et al.*, 1996.) Wischmeyer *et al.* (1997), by using the *Xenopus* oocyte system, observed that each of Kir3.2, Kir3.3, or Kir3.4 alone does not form effective functional channels. Only when Kir3.1 together with one of the other Kir3.0 members (Kir3.2, Kir3.3 or Kir3.4) are expressed, are effective channels obtained.

Single channel recordings show that expression of Kir3.2 or Kir3.4 alone produces short-duration openings (~ 0.1 ms). By contrast, expression of Kir3.1/Kir3.2 or Kir3.1/Kir3.4 result in a mixture of short (~0.1 ms) and long-duration (1-4 ms) channel openings, a characteristic gating behavior in native K_{ACh} channels (Krapivinsky *et al.*, 1995a; Kofuji *et al.*, 1995; Velimirovic *et al.*, 1996).

More recently, Jelacic *et al.* (1999) studied Kir3.3 in detail. A combination of Kir3.1/Kir3.2, Kir3.1/Kir3.4 or Kir3.1/Kir3.3 is expressed in mammalian cell lines. All of these heteromultimers produce nearly the same elementary conductance (slope conductance 35-39 pS; $[K]_o$ = 140 mM) and the same open duration (1.3-1.4 ms). On the other hand, the Kir3.2/Kir3.3 combination produces channels that are ~5 fold less sensitive to Gβγ activation than the Kir3.1/Kir3.3 combination. Channels formed by the combination of Kir3.2/Kir3.3 seem to exist in central neurons (Jelacic *et al.*, 2000).

The overall conclusion from the experiments in this section is that native Kir3.0 channels are heteromultimers.

Functional channels are not expressed if only Kir3.1 cDNA is delivered in "mammalian cells" (unlike *Xenopus* oocytes, the mammalian systems lack native Kir3.0s). Kir3.4 knockout mice do not develop K_{ACh} current (Wickman *et al*, 1998). The mechanism of this inability of functional expression is quite unique for Kir3.1. Kennedy *et al.* (1999) conducted a study on maturation of the Kir3.1 molecule. An epitope tag is attached on Kir3.1 at an extracellular domain, and the localization of the Kir3.1 protein, in relationship with the presence of the Kir3.4 protein, is observed by using the immunofluorescence technique. The study shows Kir3.1 alone, without Kir3.4, cannot be trafficked to the cell membrane. Kir3.1 can be fully glycosylated, and is properly trafficked to the cell membrane only in the presence of Kir3.4.

Are the Kir3.0 channels hetero*tetramers*? A study with electrophoresis and anti-Kir3.1 antibody demonstrated that Kir3.0 channels extracted from mouse forebrain membranes are probably tetramers composed of Kir3.1 and other proteins of a similar molecular size (Inanobe *et al.*, 1995a). Experiments using either different ratios of Kir3.1 and Kir3.4 cRNA or concatenated constructs of Kir3.1 and Kir3.4 reveal that a channel consisting of two Kir3.1 molecules and two Kir3.4 molecules produces the most vigorous activity. In addition, a configuration of **4-1-4-1** rather than **4-4-1-1** (here **4** means Kir3.4) results in most effective channels (Silverman *et al.*, 1996; Tucker *et al.*, 1996; see also Wischmeyer *et al.*, 1997.)

Experiments on purified K_{ACh} channels, using the technique of cross-linking K_{ACh} subunits with covalent bonds, demonstrate that the K_{ACh} channel is indeed a tetramer, composed of two Kir3.1 molecules and two Kir3.4 molecules (Corey *et al.*, 1998). The overall conclusion is that the K_{ACh} channel is probably a heterotetramer with a Kir3.1/Kir3.4 stoichiometry of 1 : 1, and these two subunits are arranged alternately.

4.4 Distribution of Kir2.0 and Kir3.0 in the Brain

In situ hybridization was used to study the expression of mRNA of Kirs in the brain (DePaoli *et al.*, 1994; Kobayashi *et al.*, 1995; Karschin, C. *et al.*, 1994, 1996). In neocortical cells, Kir2.1, Kir2.2, Kir2.3, Kir3.1, Kir3.2, and Kir3.3 transcripts all are expressed robustly. In the caudate nucleus, putamen, and the nucleus accumbens, Kir2.1, Kir2.2, and Kir2.3 mRNAs express strongly, while Kir3.1, and Kir3.2 transcripts are low or absent. The substantia nigra pars compacta shows almost no expression of Kir2.0 or Kir3.1 mRNAs, but there is strong expression in Kir3.2 and Kir3.3 transcripts. The locus coeruleus fails to express Kir2.0 mRNA, while it expresses a considerable amount of Kir3.1, Kir3.2, and Kir3.3 mRNA.

Distributions of Kir3.1 and Kir3.2 transcripts and proteins were studied (Liao *et al.*, 1996) using both immuno-cytohistochemistry and *in situ* hybridization. The distribution pattern of proteins, by and large, agrees with the distribution of the corresponding mRNAs. Investigations of membranes taken from these brain regions show that Kir3.1 and Kir3.2 proteins coimmunoprecipitate (Liao *et al.*, 1996), suggesting that Kir3.1 and Kir3.2 do make heteromultimeric channels. An immuno-cytohistochemical study shows that Kir3.4 proteins are expressed only in limited regions of the brain, such as the globus pallidus and the ventral striatum (Murer *et al.*, 1997).

4.5 Properties of Activated Kir3.0 Channels

4.5.1 Fast Bursts and Slow Bursts

Openings of Kir3.0 channels usually reveal two components: short-duration openings (a few hundred μs), and long-duration openings (1-3 ms). Long-duration openings exhibit very rapid bursts (Nachschlag) (Colquhoun and Sakmann, 1981): the burst consists of a succession of openings (duration, ~1 ms) separated by brief gaps (duration, a few hundred μs). Each succession of openings (a burst) is separated by longer real closings. Bursts are a characteristic feature of several channels, including the nicotinic ACh receptor-channel (Jackson *et al.*, 1983; Sine and Steinbach, 1984; Colquhoun and Sakmann, 1985; Sine *et al.*, 1990). Kir3.0 channels exhibit the "fast bursts" (Sakmann *et al.*, 1983; Grigg *et al.*, 1996). Fig. 11 shows bursts of endogenous Kir3.0 channels in heart pacemaker cells (Fig. 11*A*; Sakmann *et al.*, 1983); and in locus coeruleus neurones (Fig. 11*B*; Grigg *et al.*, 1996). The mean number of openings per long burst in rabbit pacemaker cells is 4.1 (Sakmann *et al.*, 1983); that in locus coeruleus neurones is 2.2 (Grigg *et al.*, 1996).

The "fast burst" of Kir3.0 channels was analyzed in terms of a kinetics of the type: *close* ↔ *close* ↔ *open* (Sakmann *et al.*, 1983; Grigg *et al.*, 1996). However, no strong evidence is available to indicate that this scheme is correct. Instead, we might equally well use the scheme of "open channel block": *close* ↔ *open* ↔ *close*.

Kir2.1 channels, unlike the Kir3.0 channel, do not exhibit "fast bursts". However, a mutation of the pore region (Q140E) of Kir2.1 results in channels with "fast bursts" (Guo and Kubo, 1998). This interesting observation suggests that "fast bursts" are phenomena originated from the P-region. These fast bursts may reflect a certain degree of instability of the open pore, as indicated by increase of a low frequency fluctuation during opening of nicotinic ACh channels (Sigworth, 1985).

In addition to the "fast bursts" described above, Kir3.0 channels exhibit "slow bursts" or "slow modal gating" on a very slow time scale (on the order of 100 ms to tens of s; Ivanova-Nikolova and Breitwieser, 1997; Luchian *et al.*, 1997; Ivanova-Nikolova *et al.*, 1998; Yakubovich *et al.*, 2000). Here, prolonged successions of openings are interrupted with long lasting quiescent periods (Fig. 11*D*). These slow bursts were also described by Sakmann *et al.* (1980) to explain desensitization of nicotinic ACh-receptors at a high agonist concentration.

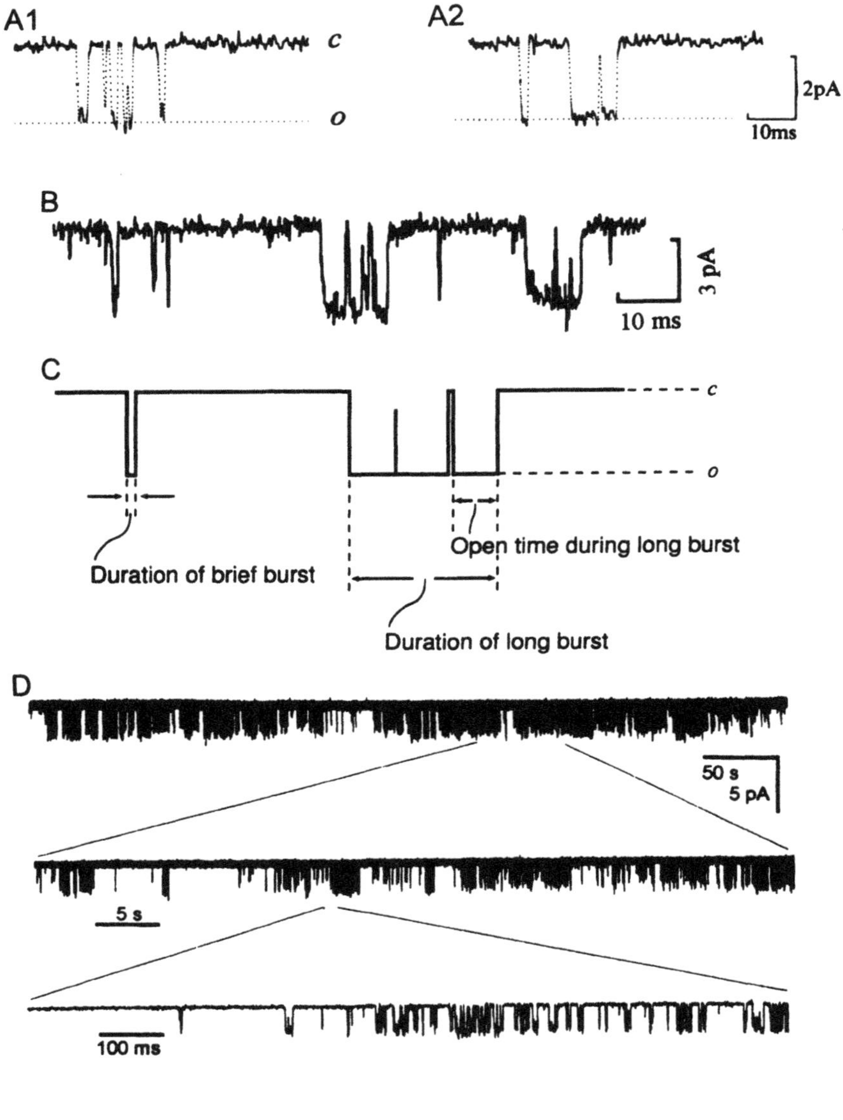
A1
A2
c
o
2pA
10ms
B
3 pA
10 ms
C
c
o
Duration of brief burst
Open time during long burst
Duration of long burst
D
50 s
5 pA
5 s
100 ms

Although the fast and slow bursts sometimes share the same terminology ("burst"), the origins of these two bursts could be entirely different from each other. Yi *et al.* (2001) and Sadja *et al* (2001) observed that mutations of various residues in the M2-region result in long trains of slow bursts. The result suggests that slow bursts could originate from the M2-region, where the channel gate is located (see also Fig. 2.3). Another possible origin is long-term changes of channel modulations ("channel gating factor"; Yakubovich *et al.*, 2000).

4.5.2 Molecular Correlates of Long and Short Openings

The above-described mode of channel opening of Kir3.0; namely, a mixture of brief openings (with a duration of a few hundred μs) and long-duration openings (1-3 ms) with bursts seems to be related to the heteromultimeric nature of Kir3.0 channels (Krapivinsky *et al.*, 1995a; Kofuji *et al.*, 1995; Velimirovic *et al.*, 1996). Heteromultimeric Kir3.0 channels consisting of Kir3.1/Kir3.2 or Kir3.1/Kir3.4 produce K^+ currents exhibiting both short and long openings, while homomultimeric channels composed of either Kir3.2 alone or Kir3.4 alone appear to give brief openings only.

Structural determinants for these different modes of Kir3.0 channel activity (brief openings or long openings) are investigated by using chimera subunits consisting of varying lengths of Kir3.4 and Kir3.1 (Chan *et al.*, 1996). Interestingly, residue F137 of the P-segment of Kir3.1 is the determinant (Table 4). Heteromultimeric channels consisting of Kir3.1, with the pore segment containing residue F137, and Kir3.4, with the pore segment containing residue S143 (residue 143 of Kir3.4 corresponds to residue 137 in Kir3.1), produce channels with short and long open times. In contrast the F137-F137 or the S143-S143 combination produces single channel activity composed of brief openings only.

Fig. 11. Burst behaviour of Kir3.0 channels. A1 and A2: Bursts of single channel currents of K_{ACh} recorded using the on-cell mode from pacemaker cells of rabbit heart. The patch pipette solution contained 70 mM K^+ and 0.2 μM ACh. External solution was a normal tyrode solution. The patch membrane was hyperpolarized from the resting potential by 20 mV. The symbols c and o, respectively, represent "closed" and "open" states. B: Bursts of single channel currents activated by somatostatin in cultured locus coeruleus neurons of the rat. The inside-out patch configuration was used. Holding potential was -95 mV. External solution (the patch pipette solution) contained somatostatin and 156 mM K^+, while the cytoplasmic side solution contained 124 mM K^+ and 0.1 mM GTP. C: This diagram illustrates various quantities used in the text and defined by Colquhoun and Sakmann (1985). D: Slow bursts (slow modal gating) of Kir3.1/Kir. 3.4 channel expressed in Xenopus oocytes. Inside-out patch activated by 20 nM Gβγ subunits. Record A is from Sakmann et al. (1983), record B from Grigg et al. (1996), and record D from Yakubovich et al. (2000). Permission obtained from Nature, J Neurophysiology, and The Physiological society. (Record A, copyright 1983, Macmillan Magazines Ltd.)

In sum, these results indicate that the P-region of Kir3.0 is a determinant for producing a mixture of long and short duration openings (F137 of Kir3.1 and the corresponding Kir3.4 residue) and burst behaviour.

4.6 Mechanisms of Rectification in Kir3.0 Channels

In Kir3.1/Kir3.2 channels as well as in the native K_{ACh} channels, removal of internal Mg^{2+} results in a milder inward rectification, with depolarizations producing substantial outward currents. Subsequent application of spermine blocks the outward currents, restoring a steep inward rectification (Horie and Irisawa, 1987, 1989; Logothetis *et al.*, 1987; Ito *et al.*, 1992; Yamada and Kurachi, 1995; Velimirovic *et al.*, 1996). These results indicate that the rectification of Kir3.0 channels originates from the same blocking mechanism (by Mg^{2+} and polyamines) as in Kir2.0. However, it is unknown whether this blocking mechanism alone is sufficient to explain all rectification in Kir3.0 channels.

Experiments by Doupnik *et al.* (1995b) are suggestive of the existence of another mechanism for Kir3.0 rectification. In Kir3.1 channels expressed in *Xenopus* oocytes, step hyperpolarizations from 0 mV induce a slow enhancement of inward current with two time constants (~ 50 ms, and ~ 400 ms). Kofuji *et al.* (1996a) analyzed this current enhancement by using chimeras between Kir3.1 and Kir1.1 (ROMK1). The Kir3.1 and Kir1.1 channels are different in two respects: (1) Kir3.1 rectification is steeper than Kir1.1 rectification, and (2) Kir3.1 channels exhibit a slow relaxation upon activation, while ROMK1 channels are activated almost instantaneously. According to Kofuji *et al.* (1996a), steepness of rectification of Kir3.1 depends on the transmembrane domains, while the slow time-dependent activation is determined by F137 in the P-region of Kir3.1 (the same residue responsible for the long open time of Kir3.1 channels, explained in the previous section). In contrast, as mentioned in earlier sections, for the Kir2.1 channel, the M2 segment and C-terminus are responsible for both the time-dependent and steep rectification. These results indicate that the slow activation of Kir3.0 and that of Kir2.0 may have different origins, suggesting the possibility that the time-dependent activation of Kir3.0 may partly originate from an "intrinsic gate" (Kofuji *et al.*, 1996a).

4.7 The Gate of Kir3.0 Channels

Taking advantage of a mutant strain of yeast, Yi *et al.* (2001) and Sadje *et al* (2001) identified probable locations of the gate of Kir3 channels. This mutant yeast, lacking the K^+ transporters TRK1 and TRK2, does not grow in low K^+ media; growth is rescued if constitutively active K^+ channels are expressed in the yeast (Minor *et al.*, 1999). Yi *et al.* (2001), using this technique, selected constitutively active Kir3.2 channels from a randomly mutated Kir3.2 library. In this way they found the following constitutively active Kir3.2: N94H, E152D, S177T, V188A, and V188G. All of these mutants show channels that are constitutively active with a long open time and slow bursts. When a hydrophobic amino acid (L, F, or I) substitutes the residue V188, the channel is not constitutively active. Substitution with any of the other amino acids (namely, small, polar, acidic, or basic amino acid) results in constitutively active channels. These observations suggested the following model: when the gate opens, the M2 segment, together with the M1 segment, rotate, resulting in an orientation in which the side chain of residue 188 occupies a pore-lining position, and any amino acids can be accommodated in the pore-lining space. When it closes, the side chain faces the interior of the channel protein, so that only some hydrophobic residues can be accommodated. Therefore, when substituted with other amino acids, the channel cannot go back easily to the closed position, resulting in a constitutively active channels (Yi *et al.* 2001; see also Fig. 3, which also summarises these experiments).

By using the same yeast screening method, Sadja *et al* (2001) also identified constitutively active Kir3.1 mutant channels. Two Kir3.1 mutants on the M2 segment are constitutively active: S170P (located near the middle of M2) and C179R (located at the end of M2). Heterotetramers consisting of these mutants Kir3.1/Kir3.4 were studied. Single channel analysis shows that the burst durations are prolonged compared to the wild type, while the single channel conductance is not altered, suggesting that the channel gate is located on the M2 region. The mutation seems to keep the gate open for a long time (with a long burst time). There seems to be a coupling between the two sites: one at the gate on the M2 segment, and the other at the P-region. The results of these two papers (Yi et al., 2001; Sadja et al., 2001) strongly suggest that the gate is located on the M2 region, and shed light on the gating mechanism.

4.8 Mechanisms of Gβγ-Kir3.0 Interaction

4.8.1 Gβγ Molecule Binds Kir3.0

Investigations up to 1995 had determined that Gβγ activates Kir3.0 channels. The next important step was to know whether this activation is due to direct interaction of the two proteins (Gβγ and Kir3.0) or is transduced via another intermediary. Co-precipitation experiments by Huang *et al.* (1995), Inanobe *et al.* (1995b), Kunkel and Peralta (1995), Krapivinsky *et al.* (1995a, b), and Doupnik *et al.* (1996) provide evidence for direct binding of Gβγ with Kir3.0 but not with Kir2.0.

Which parts of Kir3.0 molecules are the binding sites of Gβγ? Experiments using GST fusion proteins of various parts of Kir3.1 indicated that Gβγ (recombinant $G\beta_1\gamma_2$) binds a region of the C terminus (residues 273-462) and the N-terminus (1-85) (Huang *et al.*, 1995) (Fig. 2B). Gα-GDP, presumably by sequestering Gβγ, abolishes Gβγ binding to the C-terminal domain of Kir3.0. However, Gα-GDP does not sequester the Gβγ attached to the N-terminal of Kir3.1, suggesting that the heterotrimer Gαβγ, rather than Gβγ, binds the N-terminus (Huang *et al.*, 1995). Additional experiments (Huang *et al.*, 1997) indicated that Gβγ binds residue 34-86 of Kir3.1 N-terminus. Also Gβγ binds the C-terminus at residues 318-374 and 390-462 (Fig. 2B). According to Kunkel and Peralta (1995), Gβγ (recombinant $G\beta_1\gamma_2$) associates with the fusion protein of the N-terminal region (1-84), in agreement with Huang *et al.* (1995); however, Gβγ binds the initial half (182-357), not the last half (357-501), of the Kir3.1 C-terminus, in disagreement with Huang *et al.* (1997).

These results, despite some disagreements, suggest that Gβγ activates the Kir3.0 channel directly rather than indirectly via other transducers.

4.8.2 Which Kir3.0 Domains Functionally Interact with Gβγ?

Although Gβγ binds the N-terminus and parts of the C-terminus of Kir3.1 (see the previous section), binding alone does not necessarily imply channel activation. Thus, the question discussed here is: which domains of Kir3.0 interact with Gβγ and lead to channel activation? The Kir3.1 channel is activated by the interaction of the carboxyl terminus with Gβγ (Takao *et al.*, 1994). Functional tests of chimeras between Kir3.1 and Kir2.1, expressed in *Xenopus* oocytes, (Slesinger *et al.*, 1995), show that the presence of the distal part of the Kir3.1 C-terminus is sufficient for the channel merely to respond to Gβγ. However, for the channel to be activated via the receptor, both the hydrophobic core region (M1-P-M2) and the distal part of the C-terminus of Kir3.1 are necessary. Furthermore, for inducing a <u>rapid</u> response of the Kir3.0 channels upon receptor activation, all three components of Kir3.1 (the N-terminus, the hydrophobic region, and the distal part of the C-terminus) are necessary.

With regard to the mechanism of Gβγ activation of the Kir3.0 channel, an interesting notion was proposed by Luchian *et al.* (1997). Application, using the inside-out patch, of a 18-residue peptide with a sequence approximately corresponding to the end of the Kir3.1 C-terminus suppresses the Kir3.1/Kir3.4 channel activity induced by GTPγS. Interestingly, the effective concentration of the peptide is rather low (IC_{50}, 1.7 μM). Their explanation is that the C-terminus of Kir3.0 behaves like a blocking "ball", and the Kir3.0 channel is usually self-blocked. The channel opens when Gβγ couples to the Kir3.0 C-terminus, resulting in sequestering the ball, unblocking of the channel pore (see also Dascal *et al.*, 1995). This idea does not necessarily agree with an observation by Chan *et al.* (1997): they observed that a constitutively active channel does not occur after deletion of the distal part of the Kir3.1 C-terminus. However, according to Luchian *et al.* (1997), Kir3.1 C-terminal deletion constructs cannot be expressed in *Xenopus* oocytes.

Krapivinsky *et al.* (1998a) proposed a different idea, emphasizing the role of Kir3.4 rather than Kir3.1 in the activation of Kir3.1/Kir3.4 channels. Their data indicate that the beginning part of the C-terminal region (residues 209 to 225 or 226 to 245) of Kir3.4 (not Kir3.1) is crucial for the binding of Gβγ to Kir3.0 channels. A synthetic peptide corresponding to Kir3.4 residues 209-225 inhibits Gβγ binding to the K_{ACh} channel with a very high affinity (IC_{50}, 0.6 μM). Mutation of Kir3.4 at residue 216 (C216T) results in a channel that responds poorly to Gβγ. Interestingly, residue C216 is located in the C-terminal region near the inner mouth of the channel. Based on these and other results, Krapivinsky *et al.* (1998a) put forward a hypothesis that the binding of Gβγ to Kir3.4, rather than to Kir3.1, may be more critical for the activation of the channels.

Chimera constructs made of residues from Kir2.1 and Kir3.4 were tested (He *et al.*, 1999). When the segment between M323 to Y348 of the Kir3.4 C-terminus was replaced with the corresponding segment of Kir2.1, the Kir3.4 lost the capability of responding to receptor activation. He *et al.* (1999) narrowed down the residue responsible for this functional impairment, and found that L339 (of Kir3.4) is the spot critical for the interaction of GIRK with Gβγ. The corresponding residue in Kir3.1 (L333) was also critical for the sensitivity of Kir3.1 to receptor activation. Interestingly, the 'basal activity' of Kir3.0 channels, as opposed to the 'receptor-mediated activity', was not impaired by the mutation, suggesting the possibility of two different sites on Kir3.0: one that receives the Gβγ signal with high affinity (basal activity), and the other with low affinity (receptor-mediated).

According to Kennedy *et al.* (1999), Kir3.1 C-terminal domain (373-501 of Kir3.1) is not necessary for activation. A deletion mutant of Kir3.1 (delta 374-501) shows no channel activity; however, combination of mutant Kir3.1 (delta 374-501) plus wild type Kir3.4 produces normal channel activity (Kennedy *et al.*, 1999). This appears to be in conflict with the above-described results of Huang *et al.* (1995) that the Kir3.1 C-terminus is where Gβγ is attached.

These two sets of results, however, could indicate that Gβγ attaches to the Kir3.1 C-terminus or the Kir3.4 C-terminus, and attachment of Gβγ to either one of these sites is enough to activate the channels.

4.8.3 Which Gβγ Sites Interact with Kir3.0?

The quaternary structure of Gβ, elucidated through X-ray crystallography, indicates that the main part of Gβ is a seven-blade-β-propeller (Wall *et al.*, 1995; Lambright *et al.*, 1996; Sondek *et al.*, 1996). Each blade of the β-propeller consists of four anti-parallel sheets. Given these structural details, an important problem is to determine which residues of the Gβγ molecule interact with the Kir3.0. Generally Gβγ can interact with effectors only if Gα is dissociated from Gβγ; in a way, Gα acts as if it is a negative regulator for the Gβ/effector interaction. Thus, it has been imagined that Gβγ would interact with effectors at the sites that also interact with Gα. Probably based on this hypothesis, Li *et al.* (1998) and Ford *et al.*, (1998) created mutations of $G\beta_1$ at residues on the Gβ/Gα interaction sites. Then, the locations of mutations that impair the Gβ/effector interaction were determined for several different effectors (including Kir3.0 in Ford *et al.*, 1998). Ford *et al.* (1998) indicate that activation of Kir3.0 channels is impaired in mutations at the following $G\beta_1$ residues: L55 (blade 1), K78 (blade 1), I80 (blade 1), K89 (blade 1), W99 (blade 2) and D228 (blade 5). Interactions of the mutants on other effectors (PLC-β2, adenylyl cyclase I and II, etc) were also tested (Li *et al.*, 1998; Ford *et al.*, 1998). The results show that a specific interacting domain (over the Gβ surface) exists for each effector, each of these domains partially overlapping with the domains of other effectors.

Li et al. (1988) and Ford et al. (1988) investigated locations only within Gβ/Gα interaction sites, and therefore the results do not exclude the possibility that Gβ also interacts with Kir3.0 outside the Gβ/Gα interface. Albsoul-Younes et al. (2001) made mutated recombinant $G\beta_1$ proteins and used inside-out patches from locus coeruleus neurons to test their ability to activate Kir3.0 channels. Eight mutant $G\beta_1$s were made; these were chimeras in which residues located near the outermost strand of each blade were replaced with those of yeast Gβ. Yeast Gβ was assumed to be ineffective, correctly as it turned out (Peng *et al.*, 2000) in activating the K channels. The results show that the chimeras where yeast Gβ replaced parts of blade 1 and 2 (both near the outer-most strand) are unable to activate the channels, whereas these same chimeras show normal capability of interacting with adenylyl cyclases. The mutated residues of these chimeras are not located on the Gβ/Gα interaction sites, suggesting that the Gβ/Kir3.0 interaction occurs outside the Gβ/Gα interaction sites. In addition, Albsoul-Younes et al. (2001) found that W332 (blade 7) and D246 (blade 5), both of which are located on the interaction sites between Gβ/Gα, are also involved in the activation of Kir3.0 channels.

One interesting fact about the Gβ/Kir3.0 interaction is that once Gβγ starts to activate Kir3.0 channels, the activation cannot be reversed by washing Gβγ for up to ~10 minutes (Logothetis *et al.* 1987) or for as long as 30 minutes (Nakajima et al., 1996). Yet application of $G\alpha i_2$ (GDP-bound) easily stops channel activity (Logothetis et al., 1988). It is possible that Gβγ is trapped at the membrane, unable to be washed away, or that the interaction between Gβγ and Kir3.0 is quite strong and spontaneous dissociation hardly takes place. The fact that Gβ might interact with Kir3.0 at many places is in agreement with the firm attachment between Gβ and Kir3.0.

How does $G\alpha i_2$ achieve the task of reversing this firm attachment? It is known that, while the conformation of Gα changes when Gα interacts with

Gβ, conformational changes of Gβ do not occur (Lambright et al, 1996). Therefore, there has been a tendency to believe that the conformation of Gβ would not change during the interaction with effectors either. However, recent work by Loew *et al.* (1998) indicates that when phosducin interacts with Gβ, a conformation change of Gβ takes place. It is possible, as Albsoul-Younes et al. (2001) proposed, that a conformation change also occurs in Gβ, when it interacts with effectors.

4.8.4 Gγ and Kir3.0 Channels

The Gγ subunit undergoes post-translational modifications of the C-terminus, resulting in the attachment of a lipid moiety. The main step of modification is prenylation. During this step, $G\gamma_1$ (retinal γ subunits), $G\gamma_8$, and $G\gamma_{11}$ are farnesylated (15-carbon lipid moiety), whereas the majority of Gγs are geranyl geranylated (20-carbon lipid moiety) (Mumby *et al.*, 1990; Casey, 1994; Clapham and Neer, 1997). Prenylation of Gγ is needed for the regulation of adenylate cyclases (Iñiguez-Lluhi *et al.*, 1992), presumably because prenylation enables Gβγ to associate with the cell membrane (Simonds *et al*, 1991; Muntz *et al.*, 1992). A point mutation from cysteine to serine (γ2C68S) of the $G\gamma_2$ subunit at the fourth position from the carboxyl terminus blocks prenylation of γ subunits, but does not prevent the formation of a βγ complex (Muntz *et al.*, 1992). Application of β1γ2C68S to inside-out membrane patches fails to activate brain Kir3.0 channels (Nakajima *et al.*, 1996) as well as Kir3.1 channels expressed in *Xenopus* oocytes (Schreibmayer *et al.*, 1996), suggesting that anchoring of βγ subunits to the cell membrane is a prerequisite for Kir3.0 channel activation.

4.8.5 Regulators of G Protein Signalling (RGS)

Regulators of G protein signalling (**RGS**) are a family of proteins having a **GAP** (GTPase-activating protein) function, enhancing the GTPase activity of the G protein α-subunits (Berman *et al.*, 1996). In 1988, Breitwieser and Szabo observed an interesting phenomenon regarding the kinetics of K_{ACh} of atrial myocytes; the rate constant of GTP hydrolysis calculated from electrophysiological experiments was two orders of magnitude higher than that of the biochemically determined rate constant. Breitwieser and Szabo (1988) mentioned the possibility that *in vivo* K_{ACh} could be acting as a GAP. A decade later, it has been revealed that the discrepancy between the biochemical and physiological measurements originates from the presence of an RGS protein that accelerates GTP hydrolysis by $G\alpha_i$.

The functions of RGS in relation to Kir3.0 activity were investigated by heterologously expressing RGS (RGS1, RGS3, RGS4, or RGS8) and Kir3.0

channels. As expected, RGS accelerates the activation and deactivation kinetics of Kir3.0 currents (Doupnik et al., 1997; Saitoh et al., 1997; Bünemann and Hosey, 1998; Chuang et al., 1998; Herlitze et al., 1999; Kovoor et al., 2000). The apparent acceleration of activation kinetics could have been a result of acceleration of deactivation, since activation and deactivation are part of the same kinetic process (Doupnik et al., 1997; Kovoor et al., 2000). A puzzling result has remained unanswered. With a large enhancement of deactivation rate by RGS, we expect the steady-state magnitude of Kir3.0 current to be substantially decreased. In the experiments described above, the currents were not substantially decreased by the introduction of RGS.

Recently, this puzzle has begun to be solved by studies on GAP activity that the effector PLC-β1 exerts on Gαq (Posner et al., 1999; Ross and Wilkie, 2000). According to Ross and Wilkie (2000), in addition to the classical GTP-GDP exchange cycle involving G protein, an additional GTP-GDP exchange cycle starts to operate in the presence of GAP. This new signal transduction cycle involves three entities – receptor, G protein and GAP. A complex consisting of these three hydrolyses GTP quickly with all three components remaining associated. In the absence of GAP, the activated G protein is dissociated from the receptor, and after GTP hydrolysis, the re-association of the G protein with the receptor, which is the rate-limiting step, will take place. In contrast, in the presence of GAP, G proteins do not dissociate from the receptor, and the GTP-GDP exchange is greatly accelerated during activation by the agonist. Thus, the whole exchange cycle is accelerated, not just the GTP hydrolysis step. In fact, the rate constant of activation, as well as the rate constant of deactivation could be increased. It seems that these two GTP-GDP cycles can explain why acceleration of GTP hydrolysis did not result in substantial lowering of the steady-state level of activated Gα (and therefore free Gβγ) (Ross and Wilkie, 2000). Future investigations will determine whether this kind of scheme can explain the puzzle for RGS proteins, and not just for the GAP action of PLC-β1. (The functions of the Gγ subunit-like domain of RGS proteins will be treated later.)

4.9 Specificity of Signal Transduction in Kir3.0 Activation

What is the mechanism that provides the specificity in the signalling pathway arriving at Kir3.0 channels? How does a particular external signal reach Kir3.0 without cross-talk with other signals? What is the mechanism by which ACh activates K_{ACh} channels through Gβγ, while norepinephrine does not cause channel activation, even though the Gβγ liberated by the activation of the β-adrenergic receptor is capable of activating Kir3.0 channel (Lim *et al.*, 1995)? In locus coeruleus neurones, how does somatostatin open Kir3.0-like channels (see later), while substance P closes the same channels, despite release of Gβγ in both cases (Velimirovic *et al.*, 1995)?

In general terms, specificity of signalling arises from the selectivity of protein-protein interaction at each stage of the signalling pathway. There are three mechanisms that can generate specificity in influencing the Kir3.0 activity: (1) interaction between G protein and Kir3.0, (2) interaction between the receptor and the G protein, and (3) physical proximity of each component.

4.9.1 Specificity: Interaction of G Protein and Kir3.0

It is established that Gβγ, but not Gα, interacts with Kir3.0. Now many types of Gβ and Gγ have been found. Various combinations of $G\beta_1$ and $G\beta_2$ with $G\gamma_1$, $G\gamma_2$, $G\gamma_5$, and $G\gamma_7$ have been tested for their potency to activate $K_{ACh.}$ The data indicate that their potencies are not very different from each other (except for $G\beta_1\gamma_1$; Wickman *et al.*, 1994). One may conclude from this result that most combinations are equally effective, and hence the interaction between Gβγ and Kir3.0 does not generate much specificity in the signalling. This study, however, tested only $G\beta_1$ and $G\beta_2$ (in combination with 4 different Gγ's). Now, six different Gβ's (Clapham and Neer, 1997; Fletcher *et al.*, 1998) and 12 different Gγ's (Morishita *et al.*, 1995) are known, and more are likely to be discovered. It is likely that in the future Gβγ's will be regarded more important for determining the specificity of signal transduction pathways.

4.9.2 Specificity: Association of Receptor and G Protein

A specific type of receptor couples to specific types of Gα or specific combinations of Gα, Gβ, and Gγ. Excitatory receptors such as β-adrenergic receptors couple to Gαs, while inhibitory receptors couple to Gαi. Rhodopsin couples to Gt (transducin) (Freissmuth *et al.*, 1989). Experiments using antibodies and antisense oligonucleotides demonstrate that $G\alpha i_2$ (but not $G\alpha i_1$, $G\alpha i_3$, or Gαo) transduces the somatostatin-induced activation of Kir3.0-like channels in locus coeruleus neurones (Takano *et al.*, 1997). Because Gβγ is the primary agent acting on Kir3.0, the specificity of $G\alpha i_2$ is ascribed to the coupling between the receptor and the G protein subunit (Takano *et al.*, 1997). Studies of voltage-gated Ca^{2+} channels reveal that specific combinations of G protein subunits couple to specific receptors. Kleuss *et al.* (1991, 1992, 1993) investigated the signal pathways of cholinergic inhibition (via muscarinic m4 receptors) and somatostatin-induced inhibition of the L-type Ca^{2+} channel in pituitary GH_3 cells. Injecting antisense oligonucleotides into the nuclei of GH_3 cells reveals that the carbachol effect is transduced through the combination of $G\alpha o_1$/ $G\beta_3$/$G\gamma_4$ (not via $G\alpha o_2$, nor any Gαi, or any Gαs, and not via $G\beta_1$, $G\beta_2$, nor $G\beta_4$). The signal from somatostatin is transmitted through the combination of $G\alpha o_2$/$G\beta_1$/$G\gamma_3$. Their interpretation is that "the receptor recognizes and binds individual G protein subtypes, containing different β- and γ-subunits" (Kleuss *et al.*, 1992).

Various $G\beta\gamma_2$ dimers are tested for the association with recombinant Gα proteins (Fletcher *et al.*, 1998). The dimers $G\beta_1\gamma_2$ and $G\beta_2\gamma_2$ bind all types of Gα tested ($G\alpha i_1$, $G\alpha i_2$, Gαo, Gαs, and Gαq), while $G\beta_5\gamma_2$ associates only with Gαq (not with $G\alpha i_1$, $G\alpha i_2$, Gαo, or Gαs). (For the association between different types of Gβ, Gγ, and the receptor, see also Schmidt *et al.*, 1992; Pronin and Gautam, 1992; Kisselev and Gautam, 1993; and Yasuda *et al.*, 1996).

All in all, it appears that the receptors couple to G proteins with discrimination; furthermore, some types of Gβγ preferentially associate with certain types of Gα. These mechanisms would provide a substantial degree of specificity at the level of interaction among the receptors, Gα, and Gβγ.

4.9.3 Specificity: Physical Proximity

Physical proximity of the three components (receptor, G protein and Kir3.0) connected together as a compartment may also be an important mechanism for providing specificity. Compartmentalization may occur as the result of specific association of the participating proteins with each other. It is also likely to occur because of their linkage to anchoring proteins (Pawson and Scott, 1997), though most Kir3.0 channels lack PDZ domain recognition sequences (see Section 3). More anchoring proteins may be discovered in the future; therefore, as far as Kir3.0 is concerned, some, as yet unknown, protein may play a role.

(a) Lateral Mobility of Membrane Proteins, the Density of Kir Channels, and the Time Course of Receptor-Induced Kir3.0 Currents

In reconstituted lipid vesicles membrane proteins are very mobile (Neubig, 1994). Most membrane proteins in living cells, however, move slowly ($D = 10^{-10}$ $cm^2\ s^{-1}$). Many proteins, such as the subsynaptic nicotinic ACh receptors and some of the Na^+ channels, are virtually immobile, and often form aggregates (Peters, 1981; Axelrod, 1983; Almers *et al.*, 1983; Hille, 1992; Neubig, 1994).

What is the density of Kir3.0 channels over the surface membrane? Two different methods have been used to estimate the density of Kir2.0 and Kir3.0 channels (Hille, 1992) (Table 5). One of them (designated as *UC*, unitary conductance; Hille, 1992) is derived from the ratio between the maximum whole-cell conductance and the single channel conductance. This channel density (*UC*), which is relevant to our present discussion, represents the mean number of channels that are open at any moment during maximum activation of these channels in the cell. The channel density (*UC*) of Kir3.0-like channels in locus coeruleus neurones is on the order of $0.2/\mu m^2$ (Table 5).

What are the time courses of receptor-induced Kir3.0 activation? Relevant information is the time course when G protein-coupled receptors are activated suddenly and *submaximally* with an agonist at a constant concentration.

Table 5. Density of Kir2.0 and Kir3.0 channels

	temp (°C)	$[K]_o$ (mM)	γ (pS)	method	density (n /µm²)	references
Kir2.0 (guinea pig ventricles)	19-23	145	27	*PA*	0.56	Sakmann & Trube (1984)
Kir2.0 (rat skeletal muscle)	19-22	150	26	*PA*	0.09-0.36	Matsuda & Stanfield (1989)
Kir2.0 (tunicate egg)	15	50-100	5 - 6	*UC*	0.03 - 0.04	Ohmori (1978); Fukushima (1982)
Kir3.0 (locus coeruleus)	31	10	30*	*UC*	0.2*	Grigg et al.(1996); Velimirovic et al. (1995)

Methods of calculation: *PA* (patch area method): the number of channels in a patch is divided by an estimated patch area. This will give a number of channels that can be activatable (even though some are closed at any given time). *UC* (unitary conductance method): the maximum conductance in whole-cell experiments divided by the unitary conductance (the same method as in *UC* in Hille, 1992; page 330, table 4). The *UC* method will give the average number of channels that happen to be open at any instant during the maximum activation of whole-cell Kir current.

* The value of γ in Grigg et al. (1996) was obtained at 23°C with $[K]_o$ = 156 mM and 50 mM. The value of γ in this table (31°C; $[K]_o$ = 10 mM) was calculated by taking a temperature coefficient of γ being 1.7 (Sakmann and Trube; 1984) and γ being proportional to $[K]_o$ with an exponent of 0.18 (from the data of Grigg et al., 1996).

Examples in Fig. 12 (*A*1 and *B*1) show that the responses at submaximal agonist concentrations reach a plateau 5-10 s after application of agonists.

From these data can we answer the question whether these time courses of Kir3.0 activation are compatible with the diffusion of Gβγ from the location of the receptor to the sites of the channels? Suppose that a conglomeration of the receptors forms a long rectangular source of Gβγ proteins and that upon stimulation of the receptors a steady-state concentration of Gβγ subunit is set up and maintained at this source. If Kir3.0 channels are located on the average 1 μm from the receptor with D = 10^{-10} cm^2 s^{-1}, the Gβγ concentration at the site of Kir3.0 channels at 10 s would be only ~2.5 % of the concentration of the source (Crank, 1967; Eq. 2.45). If the channels are located at 0.1 μm from the source, then at 10 sec, the concentration reaches almost 80 % of the original concentration. This seems to support the view that some compartmentalization is necessary for effective transduction. However, the answer depends very much on the value of D. If we take a larger value of D (= 10^{-9} cm^2 s^{-1}), the Gβγ concentration at 1μm at 10 s would become ~ 50 %; howeve, the Gβγ concentraion at the channel location is still being built up. Therefore, it would be difficult for the receptor stimuation to set up the observed depolarizations (Fig. 12) without assuming compartmentalization.

(b) Evidence for Compartmentalization

As mentioned in Section 4.8 (Gβγ-Kir3.0 interaction), Gβγ binds to the C-terminal domain of Kir3.1, while the trimeric G protein (Gαβγ) associates with the N-terminus of Kir3.1 (Huang *et al.*, 1995). Physiological experiments using various chimeras of Kir3.1 and Kir2.1 (Slesinger *et al.*, 1995) show that in the absence of the Kir3.1 N-terminus, receptor-induced responses of Kir3.0 channels become slow. Based on these results, Huang *et al.* (1995) and Slesinger *et al.* (1995) suggest the presence of compartmentalization consisting of the receptor, the G protein, and the Kir3.0 channel, the association of Gαβγ and the Kir3.1 N-terminus helping the three components to assemble.

The mobility of G protein subunits was determined by Kwon *et al.* (1994) by measuring the fluorescence recovery after photo bleaching. Only 16 % of the total Gβγ is rapidly mobile (D = 2 x 10^{-9} cm^2 s^{-1}), whereas the remaining Gβγ (84 %) is virtually immobile. This observation favors compartmentalization.
It is known that Gα monomers associate with tubulin (Roychowdhury *et al.*, 1993), and Gβγ dimers associate with microtubules (Roychowdhury and Rasenick, 1997) and with actin filaments (Carlson *et al.*, 1986). There are several morphological investigations regarding the pattern of localization of Gβγ. In NG-108-15 cells confocal microscopy shows a punctate pattern of Gβγ localization, suggesting assemblies of these proteins (Kwon *et al.*, 1994). Distribution of Gβγ on isolated plasma membranes from COS cells transfected with Gβγ also indicates a punctate pattern (Muntz *et al.*, 1992). Immunocytochemistry in cardiac fibroblasts reveals that $G\gamma_5$ colocalizes with vinculin and stress fibres (Hansen *et al.*, 1994).

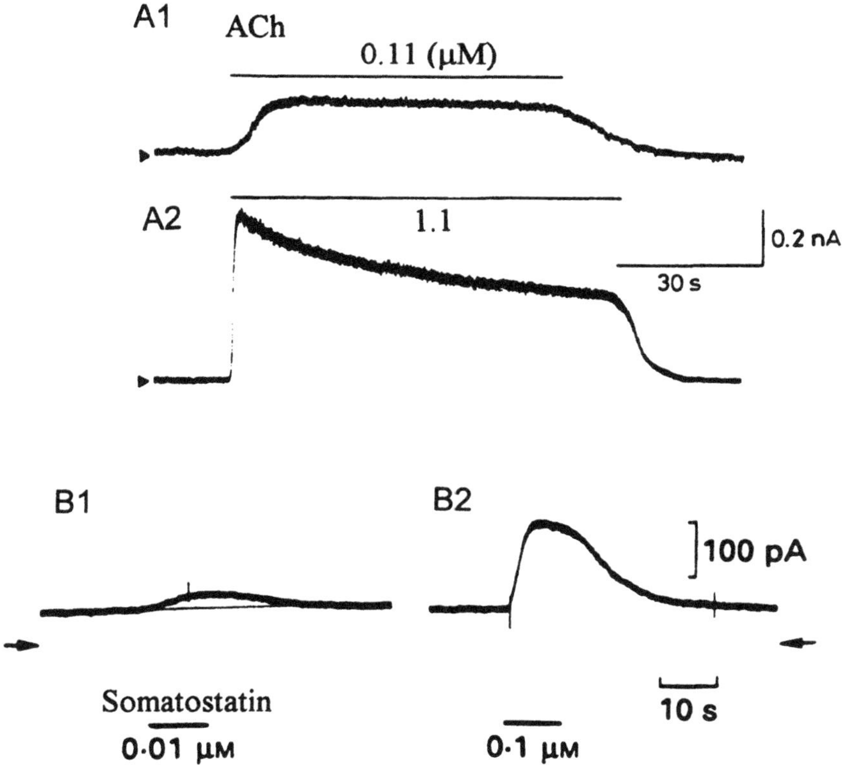

Fig. 12. Time courses of transmitter activated Kir3.0 currents. A: Kir3.0 currents induced by ACh in atrial cells of the guinea pig heart by using the whole-cell recording. The membrane potential was held at -53 mV. Two different concentrations (0.11 μM and 1.1 μM) of ACh were applied during the time indicated by the horizontal bar. B: Somatostation-induced current in cultured locus coeruleus neurones. Holding potential was -80 mV. Somatostation was applied by pressure ejection through a glass pipette. Application of a high K^+ solution showed that the on halftime of the concentration change was 0.9 s using this pressure ejection method. A is from Kurachi et al. (1987), and B is from Inoue et al. (1988). Permissions obtained. Copyright in A is in Springer-Verlag; B is in the Physiological Society.

In summary, results from experiments with different approaches strongly suggest the existence of compartmentalization consisting of the receptors, G proteins, and Kir3.0. However, crucial evidence about the compartmentalization is lacking.

5 Modulators of Kir3.0

In this section, we deal with modulation of the Kir3.0 channels. In some instances "gating" and "modulation" are distinctly different phenomena. In voltage-gated K^+ channels a change of membrane potential is the primary agent that opens or closes the channel gate; other factors influencing the channel functions can be regarded as modifiers or modulators, and no ambiguity remains. In Kir channels, however, the distinction is less clear. Here, we will use the term modulation to mean any modifying influences, except for the regulation by voltage changes, on the functions of Kir3.0. General properties of modulation by Gβγ have already treated.

5.1 PIP_2 and ATP

We have already discussed the general properties of the interaction between phosphatidyl inositol 4,5-bisphosphate (PIP_2) and Kir channels (see Section 3.4). Here we discuss experiments concerned with the way in which PIP_2 and Gβγ interact in Kir3.0 channel gating.

Huang *et al.* (1998) analyzed the roles played by PIP_2 in the regulation of the Kir channel activity and concluded that 'the GIRK channel can be activated by PIP_2 without the presence of Gβγ' (even after sequestering Gβγ by Gαi). They proposed (Huang *et al.*; 1998) that 'Gβγ activates GIRK channels by stabilizing interactions between PIP_2 and the K^+ channel'.

Sui *et al.* (1998) studied the roles of PIP_2 in the regulation of Kir3.0 channels at the single channel level; they used mainly Kir3.1 and Kir3.4 heterologously expressed in *Xenopus* oocytes. To prevent rundown the presence of MgATP in the cytoplasmic side is a necessity. Rundown is prevented or reversed by applying PIP_2. Antibody against PIP_2 induces rundown. Phosphatidylinositol-specific PLC, which hydrolyzes phophatidylinositol, decreases channel activity; PLC-β2, which hydrolyzes PIP_2 to inositol triphosphate and DAG, lowers channel activity. Interestingly, application of PIP_2 as well as MgATP, not only increases the frequency of channel opening but also increases the duration of channel openings, a fact that Kim (1991) had previously observed. From these results Sui *et al.* (1998) conclude that 'phosphatidylinositol phospholipids seem to play a permissive role by regulating the effectiveness of gating molecules such as Gβγ subunits or Na^+ ions, rather than activating the channel themselves'.

By using atrial myocytes in inside-out configuration, Kim and Bang (1999) compared the effects of various phospholipids (including PIP_2). In the absence of Gβγ and Na^+, none of the phospholipids could activate the Kir3.0 channel; instead they inhibit the initiation of channel activity. However, if the Kir3.0 channel is already activated by Gβγ, the application of phospholipids enhances the channel activity and increases the duration of channel opening. Thus, phospholipids alone (without Gβγ) appear not to activate Kir3.0, but they enhance the Kir3.0 activation by Gβγ.

In sum, according to Huang *et al.* (1998), phospholipids, in the absence of Gβγ, can activate Kir3.0, and therefore Gβγ helps phospholipids to activate the channel. In contrast, according to Sui *et al.* (1998) and Kim & Bang (1999), phospholipids may be a a necessity for Kir3.0 activation, but are not sufficient; Gβγ is also required. How do these new revelations fit to the older idea? These new discoveries do not necessarily contradict the idea that Gβγ directly activates Kir3.0. Receptor activation by a transmitter would liberate Gβγ, and the increased level of Gβγ would activate Kir3.0 channels. The knowledge on the role of PIP_2 has given us a deeper understanding of the way Gbγ enhances the Kir3.0 activity.

When ATP is applied to Kir3.0 channels, it is expected that the PIP_2 level in the membrane would increase; this increase of PIP_2 level would enhance the channel activity. In contrast, Medina *et al.* (2000) observed that the K_{ACh} channel is modulated by phosphorylation and dephosphorylation of the channel (or of an accessory protein) in atrial mytocytes. Pretreatment with phosphoprotein phosphatase 2A (PP2A) prevents Gβγ from activating Kir3.0 channels, indicating that the phosphatase effectively dephosphorylates a site whose phosphorylation is a prerequisite for the channel to be activated by Gβγ. Curiously, once Gβγ initiates the Kir3.0 channel activity, application of PP2A was not capable of suppressing the channel activity. This behavior of PP2A is different from the situation of a change of PIP_2 level. Even if Kir3.0 is activated by Gβγ, hydrolysis of PIP_2, or the prevention of its function by anti-PIP_2 antibody, results in suppression of the Kir3.0 channel.

Thus, ATP seems to have two types of effect. One is to enhance the PIP_2 level; the other is to phosphorylate the channels (or an accessory protein). Both increase the activity of Kir channels. The roles of PIP_2 in the inhibition or desensitization of Kir3.0 will be treated in Section 5.3.6a.

5.2 Intracellular Na^+

Sui *et al.* (1996) observed that Kir3.0 channels are activated by millimolar concentrations of intracellular Na^+ (with an EC50 ~ 40 mM), provided that the channels are treated by Mg-ATP. Application of cardiac glycosides, which would result in an increase in $[Na^+]_i$ by inhibiting the Na^+/K^+ pump, also activated Kir3.0 channels. This effect could be partly responsible for the bradycardia caused by cardiac glycosides.

The molecular mechanism of the Kir3.0 sensitivity to Na^+ was clarified by Ho and Murrell-Lagnado (1999a, b). The amino acid residue that is responsible for this Na^+ sensitivity is D226 for Kir3.2 (D223 for Kir3.4; Zhang *et al.*, 1999). The affinity of Kir3.2 for PIP_2 is enhanced by increasing $[Na^+]_i$ or by

D226N mutation (Ho and Murrel-Lagnado, 1999a). It is likely that neutralization of the negatively charged residue (D226) by Na^+ or by the D226N mutation would enhance the affinity of Kir3.0 C-terminus to PIP_2 (which is negatively charged), resulting in the activation of Kir3.0 channels. In Kir3.2, D226 is close to the sites for PIP_2 binding (see Section 3.4). Zhang *et al* (1999) reached the same conclusion for Kir3.4, where the molecular determinant for Na^+ activation is D223 (the counterpart in Kir3.4 of D226 in Kir3.2).

5.3 Synaptic Potentials and Channel Modulation

One of the most biologically significant roles of Kir channels is to act as an effector of transmitter and hormone actions. Transmitters acting on G protein-coupled receptors (GPCRs) induce "slow synaptic potentials" lasting seconds or minutes (slow synaptic actions, Hille, 1992). In this section, we deal with the modulation of Kir channels in relation to the generation of slow synaptic potentials.

5.3.1 Slow Synaptic Potentials

Up to the early 1980s, "slow synaptic potentials" or "slow synaptic actions" (Hille, 1992) caused by transmitters were investigated in sympathetic, parasympathetic and brain neurones. The mechanisms of slow hyperpolarizations or depolarizations were mainly (but not altogether) attributed to changes in K^+ conductance (Kuba and Koketsu, 1976; Adams *et al.*, 1982; Jan and Jan, 1982; Williams *et al.*, 1982; Kuffler and Sejnowski, 1983; North and Williams, 1985; Madison *et al.*, 1987). In addition to K^+ conductance changes, certain slow synaptic potentials were generated by non-selective cation channels (Kuffler and Sejnowski, 1983; Koyano *et al.*, 1993; Farkas *et al.*, 1994, 1996; Jiang *et al.*, 1994). A discussion of the non-selective cation channels, however, will not be included here.

Depending on the materials and conditions, either the M-current, or a linear K^+ leak conductance was shown to be the K^+ conductance involved. The presence of classical inward rectifiers in neurones had been known for some time (Kandel and Tauc, 1966; Constanti and Galvan, 1983; Osmanović and Shefner, 1987). However, it was not suspected that inwardly rectifying K^+ conductance would be responsible for generating synaptic potentials.

Benson and Levitan (1983) first described the role of inward rectifier K^+ channels as effectors of neuronal synaptic action in *Aplysia* neurones. Here, serotonin was shown to induce hyperpolarization by activating an inward rectifier. This paper was followed by a report on the mammalian brain (Stanfield *et al.*, 1985), in which substance P was shown to produce a slow excitation by suppressing an inward rectifying K^+ conductance in neurones from the nucleus basalis. Next, many transmitters were reported to produce slow "hyperpolarizations" by activating Kir in various types of neurones.

5.3.2
Slow Inhibitory Synaptic Potentials in Neurones

Aghajanian and Wang (1986) and Andrade *et al.* (1986) showed that the increase in K^+ conductance produced by various slow inhibitory transmitters in brain neurones is mediated by a G protein, even though the channels were not identified as being inward rectifier K^+ channels.

Since then, many slow inhibitory transmitters have been shown to cause hyperpolarizations by activating inward rectifier K^+ channels, and the signals have been shown to be mediated by pertussis toxin sensitive G proteins. Most probably these inward rectifiers are Kir3.0 channels. A partial list of transmitters acting in this way is: adenosine (Trussel and Jackson, 1985, 1987), opioids (North *et al*, 1987; Williams *et al.*, 1988b; Wimpey and Chavkin, 1991), somatostatin (Mihara *et al.*, 1987; Inoue *et al.*, 1988), serotonin (Williams *et al.*, 1988a; Penington *et al.*, 1993a), noradrenaline (Surprenant and North, 1988), muscarinic agonists (Gerber *et al.*, 1991), dopamine D2 agonists (Kim, KM *et al.*, 1995; 1997), and GABA (Sodickson and Bean, 1996). (See reviews by North, 1989; Brown, 1990; Nakajima and Nakajima, 1994).

The signalling mechanism that produces hyperpolarization is essentially the same as that for the ACh-induced hyperpolarization in atria cells. Pertussis toxin abolishes transmitter-induced activation of Kir channels in locus coeruleus neurones, raphe dorsalis neurones, and other brain neurones (Trussell and Jackson, 1987; Inoue *et al.*, 1988; Williams *et al.*, 1988a). The activation of Kir3.0-like channels induced by adenosine, opioids, somatostatin, or α2-adrenergic agonists depends on the presence of GTP in the cytoplasm. Non-hydrolyzable GTP analogues produce sustained activation of Kir channels (Trussel and Jackson, 1985, 1987; North *et al*, 1987; Mihara *et al.*, 1987; Inoue *et al.*, 1988; Surprenant and North, 1988; Velimiromic et al., 1995). Transgenic mice devoid of Kir3.2 gene do not produce postsynaptic inhibitory potentials through $GABA_B$, 5-HT_{1A}, and adenosine A1 receptors (Lüscher et al., 1997).

5.3.3
Gβγ Activation of Neuronal Kir3.0

Single channel analyses were performed on Kir3.0-like channels in brain neurones by using the cell-attached configuration in locus coeruleus neurones (Miyake *et al.*, 1989), and the outside-out configuration in nucleus raphe dorsalis neurones (Penington *et al.*, 1993b). Oh *et al.* (1995) recorded single channels activated by serotonin in acutely dissociated hippocampal neurones.

Single channel analysis was performed, using locus coeruleus neurones with the inside-out configuration on Kir3.0-like channels that are activated by somatostatin or opioids (Grigg *et al.*, 1996). A subsequent study shows that Kir3.0-like channels from the locus coeruleus (LC) are activated by recombinant $G\beta_1\gamma_2$ applied to the cytoplasmic side of the inside-out patches (Nakajima *et al.*, 1996; Albsoul-Younes et al., 2001). Once the Kir3.0-like channels in LC are activated by $G\beta_1\gamma_2$, the activity cannot be reversed by washing away the $G\beta_1\gamma_2$; the activity is reversed only if $G\alpha i$ (GDP bound) is applied. A very high concentration (100 nM) of recombinant activated $G\alpha i_2$ does not activate the Kir3.0-like channels. Thus, as in the case of the atrial K_{ACh} or the cloned Kir3.0 (Logothetis *et al.*, 1987; Reuveny *et al.*, 1994; Wickman *et al.*, 1994), brain Kir3.0 is activated by $G\beta\gamma$, but not by $G\alpha$. Overall, the properties of the brain Kir3.0 appear to be similar to those of the K_{ACh} channels.

5.3.4 Slow Excitation in Nucleus Basalis Neurons

Substance P excites cholinergic neurones from the nucleus basalis by suppressing an inward rectifying K^+ current. This was the first demonstration of transmitter-induced "inhibition" of inward rectifier K^+ channels (Stanfield *et al.*, 1985). What are the properties of this Kir channel that is suppressed by substance P in nucleus basalis neurons? Is it one of the Kir3.0 (GIRK) family? The molecular identity of the Kir channel in the nucleus basalis is yet unknown. We call this the KirNB channel.

After nucleus basalis neurones are loaded with a non-hydrolyzable GTP analogue (GTPγS or GppNHp), application of substance P produces an irreversible suppression of the KirNB current. This substance P effect is pertussis toxin insensitive (Nakajima *et al.*, 1988). Subsequent experiments on KirNB channels (Takano *et al.*, 1995), using cell-attached single channel recordings, revealed that the KirNB channel seems to be constitutively active, and the elementary conductance (~ 20 pS at $[K^+]_o$ = 156 mM) is smaller than that of Kir3.0 channels (32 to 40 pS). The open duration of the KirNB channel is similar to that of Kir3.0 channels (~1 msec), and far shorter than the values for Kir2.0 channels (10 to 100 ms) (Takano *et al.*, 1995).

In the cell-attached mode substance P application outside the patched area suppresses channel activity, indicating that a diffusible messenger travels from the extra-patched area into the patched area. The PKC inhibitors staurosporine and PKC(19-36) (a pseudosubstrate peptide for conventional PKCs) suppress this effects of substance P. When neurones are treated with okadaic acid, the substance P-induced suppression of KirNB occurs, but the suppression does not recover, indicating that the recovery from the suppression is caused by dephosphorylation by a serine/threonine protein phosphatase (Takano *et al.*, 1995). Further experiments using antibody injection with a microinjector show that the substance P-induced suppression of the KirNB is mediated by Gq/11 and PLC-β1 (Takano *et al.*, 1996).

In summary, substance P suppresses KirNB channels in nucleus basalis neurons. Because KirNB channels are constitutively active in the resting state, a "resting" neuron can be excited by substance P. The signal transduction of the substance P effect involves activation of protein kinase C (PKC) and phosphorylation of the KirNB molecule (or the accessory protein). Dephosphorylation by a protein phosphatase results in the reopening of the KirNB channels.

There are several reports describing transmitter-induced suppression of inward rectifier K^+ currents leading to slow excitation of neurones. Serotonin and muscarine suppress Kir channels in nucleus accumbens neurones (North and Uchimura, 1989; Uchimura and North, 1990). Muscarine reduces Kir in sympathetic ganglia neurones (Wang and McKinnon, 1996). The signalling pathway of these effects has not been fully investigated but could be similar to those described in the KirNB channel.

5.3.5 Slow Excitation in Locus Coeruleus and Other Neurons

Locus coeruleus neurones are innervated by both inhibitory and excitatory neurones. Koyano *et al.* (1993) and Velimirovic *et al.* (1991,1995) investigated the interaction of an inhibitory transmitter and an excitatory transmitter on Kir3.0-like channels of locus coeruleus neurones. Locus coeruleus neurones are loaded with GTPγS; application of somatostatin (or metenkephalin) produces enhancement of an inward rectifier current (representing the activation of Kir3.0-like channels). This enhancement of Kir3.0-like current is sustained (owing to the presence of GTPγS); but application of substance P abolishes the enhancement (Fig. 13B). The larger the enhancement of the Kir3.0-like current, the greater the suppression by substance P, suggesting that the same channels that are enhanced by somatostatin are neutralized by substance P. Once substance P suppresses this current, somatostatin cannot enhance the current any more. The opening of the Kir3.0-like channel by somatostatin is mediated by a pertussis-toxin sensitive G protein, Gi_2 (Inoue *et al.*, 1988; Takano *et al.*, 1997), while the closing of the opened channel by substance P is mediated by a pertussis toxin-insensitive G protein (Velimirovic *et al.*, 1995). Thus, the Kir3.0-like channel in locus coeruleus is under a dual, opposing regulation. The same Kir3.0-like channel molecule that responds to an opening signal from the pertussis toxin-sensitive G protein (Gi) also responds to a closing signal from the pertussis-toxin insensitive G protein. The closing signal is stronger than the opening signal. Because this neutralization of the Kir3.0-like channel activity by substance P occurs in the presence of GTPγS, it cannot be attributed to sequestration of Gβγ by Gα. Although the main experiments were done in the presence of GTPγS, essen-

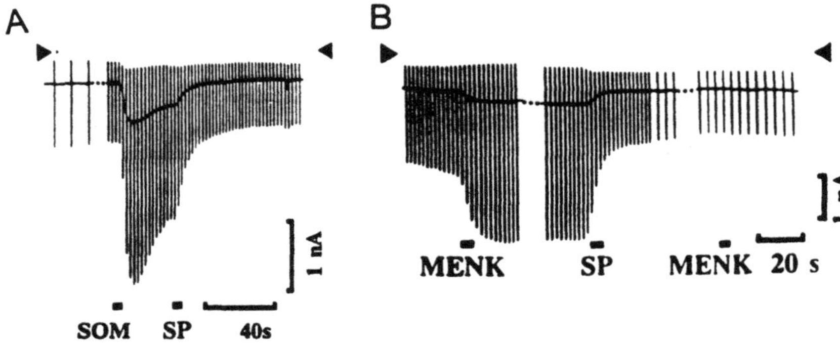

Fig. 13. Somatostatin (SOM) or metenkephalin (MENK) activates Kir3.0 (GIRK) currents whereas substance P (SP) suppresses the Kir3.0 currents in cultured locus coeruleus neurons. A: Interaction of SOM (0.2 μM) and SP (0.3 μM) effects. Whole-cell recording with a patch pipette solution containing GTP and Mg-ATP. The SP application completely eliminated the Kir3.0 current that was activated by SOM. The effects of both SP and SOM partially recoverd after a few minutes (not shown). External solution was 10 mM K^+ Krebs; holding potential, -74 mM. B: Whole-cell recording with a patch pipette containing GTPγS. External solution was a 10 mM K^+ Krebs solution. MENK activated Kir3.0 current, and SP eliminated it; thereafter, a second MENK application did not activate Kir3.0 current any more, and the effects of SP remained without recovery (because of GTPγS). Interval between the first MENK and SP application was 102 sec, and that between SP and the last MENK was 120 sec. To measure membrane conductance, command pulse sequences were periodically applied. Each command pulse sequence consisted of a square-wave depolarization (20 mV, 100 ms) and a hyperpolarization (50 mV, 100 ms); at this slow time base, each sequence appears to be a single vertical line. Arrowheads indicate zero current level. From Velimirovic et al. (1995), with permission. Copyright (1995), National Academy of Sciences, USA.

tially the same results were obtained in the presence of GTP (Fig. 13A), except that the effects were transient in the absence of GTPγS. The signal transduction pathway of this neutralization of Kir3.0-like channels is yet unknown. More recent experiments by Farkas *et al.* (1997) showed that dopamine D2 agonists open Kir3.0-like channels in brain dopaminergic neurones, whereas the excitatory transmitter neurotensin closes the Kir3.0-like channels. Therefore, this type of opposing regulation of Kir3.0-like channels could be a widespread phenomenon in brain neurones.

5.3.6 Inhibition of Kir3.0 Channels in Atrial Myocytes and Hetorologous Expression Systems

(a) PIP_2: Atrial myocytes are innervated reciprocally by the vagus and the sympathetic nerves. Braun *et al.* (1992) showed that methoxamine, an α1-

adrenergic agonist, reduced atrial Kir3.0 channel activity. In this interesting paper, Braun *et al.* (1992) further observed that the signal transduction of this Kir3.0 suppression does not involve PKC.

More recently, Kobrinsky *et al.* (2000) studied desensitization of the Kir3.0 activity in atrial myocytes and in heterologously expressed systems (COS-1 cells and Xenopus oocytes). Their results indicate that in the continuous presence of ACh, Kir3.0 channels are activated (through m2-muscarinic receptors) and then decline to a lower level within a minute (through m3-muscarinic receptors). This fairly quick decline does not involve receptor desensitization, and is called 'short-term desensitization'. The conclusion of Kobrinsky *et al.* (2000) is that the same ACh that activates the m2 receptor activates the m3 receptor. Activation of m3 receptors in turn activates Gq, resulting in activating PLC, with the consequence that membrane PIP_2 levels are reduced, causing a decline of Kir3.0 activity. In their experiments, Kobrinsky *et al.* (2000) used pharmacological agents as well as molecular biological tools. For example, they used U73122 (an inhibitor of PLC) as well as a peptide (Wu *et al.*, 1993) that traps Gαq in a dominant negative way. Another ingenious technique was using a Kir3.0 mutant that binds PIP_2 more strongly than does wild type Kir3.0. When mutant Kir3.0 was expressed, the desensitization became less, strengthening the idea that the desensitization is caused by reduction of the PIP_2 level. They also used a mutant Kir2.0, whose sensitivity for the detection of a decline in the level of PIP_2 is greater than that of wild type Kir2.0. This mutant channel could then be used as a gauge for detecting the PIP_2 level. Indeed, they detected a reduction of the PIP_2 level when ACh was applied.

Unlike Kobrinsky *et al.* (2000), who studied 'desensitization' caused by a single transmitter ACh, Cho *et al.* (2001), Lei *et al.* (2001), and Meyer *et al.* (2001) investigated the reciprocal effects of inhibitory and excitatory transmitters. Cho *et al.* (2001) analyzed inhibition of the Kir3.0 channel by the α1-adrenergic agonist phenylephrine in atrial myocytes, and concluded that the inhibition arises from a decline of the PIP_2 level, which is caused by PLC activation. Lei *et al* (2001) studied HEK 293 cells transfected with the $5\text{-}HT_{1A}$ receptor and thyrotropin-releasing hormone (TRH) receptor, and concluded that the TRH-induced inhibition of Kir3.0 channels arises from reduction of the membrane PIP_2. Meyer *et al.* (2001) also concluded that an inhibition of Kir3.0 channels in atrial myocytes by phenylephrine (α_1-adrenergic agonist) and by endothelin-1 result from reduction of the PIP_2 level in the membrane originated from the signaling pathway through Gαq/11 and PLC.

The effector for transmitter-induced excitation via lowering of PIP_2 is not confined to Kir3.0. Xie *et al.* (1999) presented evidence indicating that cloned K_{ATP} channels (Kir6.2 and SUR2A) expressed in COS-7 cells are inhibited through m1-muscarinic receptor via lowering of PIP_2. Overall, the evidence for the role of PIP_2 in transmitter-induced inhibition of Kir3.0 channels has become quite strong.

(b) PKC: Although the evidence for the role of PIP_2 in transmitter-induced inhibitions of Kir3.0 has become stronger, other results indicate that inhibition of Kir3.0 results from a PKC-induced phosphorylation of the channel protein.

Sharon *et al.*, (1997) studied inhibition of Kir3.1 currents induced by activation of metabotropic glutamate receptors by using the *Xenopus* oocyte expression system. Activation of the metabotropic glutamate receptors, known to activate Gq and PLC, inhibits Kir3.1 activity. This inhibition is blocked by the broad-spectrum protein kinase inhibitor staurosporine and by the specific PKC inhibitor bis-indolylmaleimide, but not by calphostin C, suggesting that an isoform of PKC may be the messenger of this signal.

Stevens et al. (1999) investigated the bombesin-induced inhibition of the Kir3.0 currents by using the *Xenopus* oocyte system. The Kir3.0 channels are obtained from the following combinations: Kir3.2 alone, Kir3.1/Kir3.2, and Kir3.1/3.4. Applications of the PKC inhibitors chelerythrine and staurosporine reduce the bombesin-induced inhibition of the Kir3.0 currents. Activation of PKC with phorbol esters also inhibits the Kir3.0 currents. These results suggest that Kir3.0 currents are regulated by phosphorylation through PKC. Hill and Peralta (2000) expressed Kir3.1, Kir3.4, m1-muscarinic receptor, and D2-dopamine receptor in *Xenopus* oocytes. They concluded that inhibition of Kir3.0 channels induced by the m1-muscarinic receptor is caused by PKC-induced phosphorylation.

More recently, Leaney et al. (2001) showed, by using the mammalian expression system HEK293, that the Kir3.1/3.2a currents are inhibited by m1 and m3 muscarinic receptors. Application of PKC inhibitors, staurosporine, bisindolylmaleimide I, and Ro-31-8220, reduces the muscarinic receptors-induced inhibition. Direct activation of PKC by phorbol esters or by a diacylglycerol analogue completely occludes the muscarinic receptor effect. The muscarinic receptor-induced inhibition of Kir3.0 occurs in the absence of cytosolic Ca^{2+}. Western blotting indicates the presence of conventional PKCs and novel PKCs (which are Ca^{2+}-independent). Stimulation by phorbol ester results in translocation of novel PKCs (PKCδ and PKCε) from cytosol to particulate fractions. Altogether, this work has strengthened the evidence for the role of PKC in the inhibition of Kir3.0 channels.

Summarising the evidence explained in (a) and (b), it is likely that both PKC and PIP_2 are involved in the inhibition of Kir3.0; the degree of the involvement from each of these signal pathways varies according to the cell type and the conditions.

(c) $G\beta_5$: $G\beta_5$ is a most distantly related member of Gβ family. $G\beta_5$ is suggested to be the type of β-subunit that couples to Gαq (Fletcher *et al.* 1998). Lei *et al.* (2000) reported that expression of $G\beta_5$ diminishes Kir3.0 currents (both basal and transmitter-activated Kir3.0 currents) by about half. Based on these data Lei *et al.* (2000) suggested that suppression of Kir3.0 induced by Gαq could ultimately be attributable to $G\beta_5\gamma_2$.

In contrast, data, which assign different roles of Gβ_5 have appeared (Ross and Wilkie, 2000). Snow *et al.* (1998, 1999) reported that some types of RGS (RGS 6, 7, 9, and 11) have a G protein gamma subunit-like (GGL) domain, and Gβ_5 interacts specifically with these RGSs. The Gβ_5/RGS complex could act as a GAP (GTPase-activating protein) preferentially on Gαo. Endogenous Gβ_5 is isolated as a soluble complex with RGS7 from the retina (Cabrera *et al.*, 1998). In the brain Gβ_5 *in vivo* also interacts more preferentially with GGL-containing RGSs than with Gγ (Zhang and Simonds, 2000). Kovoor *et al.*, (2000) have shown that Gβ_5 greatly enhances the GAP activity of RGS7 or RGS9, indicating that Gβ_5 works to promote the activity of certain RGSs.

5.3.7 Comparison of Kir3.0 and KirNB

Both Kir3.0 channels and KirNB channels (constitutively active Kir channels in the nucleus basalis) play important roles in determining the excitability of noradrenergic neurons and cholinergic neurons in the brain. Kir3.0 channels are activated by inhibitory transmitters, through G$\beta\gamma$, and suppressed by excitatory transmitters through a pertussis toxin insensitive G protein. The signal transduction pathway of this suppression has not been determined yet.

KirNB channels are different from Kir3.0 channels: KirNB channels are constituively active, and the elementary conductance seems to be smaller than that of Kir3.0. The activity of KirNB channels is inhibited by phosphorylation, and then the activity is recovered by dephosphorylation. It is possible that KirNB is neither Kir2.0 nor Kir3.0. KirNB has been described only in the nucleus basalis. But it is unlikely that KirNB, which is the target channel for an excitatory transmitter substance P, exists only in the nucleus basalis. They probably are located in many other nuclei of the brain, and may play an essential role in determining excitability of CNS neurones.

5.3.8 Other Examples of Kir Modulation

We have already discussed, in detail, the modulation of Kir2.0 channels (Section 3.7). In that section we describe how Koumi *et al.* (1995a,b) showed that isoproterenol reduces the activity of Kir2.0 in heart through PKA-dependent phosphorylation. In contrast, with Kir3.0, either in atrial cells or with cloned Kir3.1/Kir3.2 heterologously expressed in *Xenopus* oocytes, application of ACh together with isoproterenol enhanced the Kir3.0 currents as compared with application of ACh alone (Müllner *et al.*, 2000). This increase seems to be caused by phosphorylation through PKA downstream from the β2-adrenergic receptor.

Kir channels (perhaps Kir4.0 channels) in glia (astrocytes) are inhibited substantially by a β-adrenergic agonist (Roy and Sontheimer, 1995). This inhibition is mediated by cyclic AMP, but not through cyclic AMP-dependent protein kinase (PKA). Cyclic AMP seems to act as a ligand to reduce the Kir channel activity; in other words, the activity of the Kir is cyclic nucleotide-dependent.

In oligodendroglia, several G protein-coupled receptors, such as somatostatin receptors, serotonin receptors, and muscarinic receptors, are present (Karschin, A. *et al.*, 1994). Activation of these receptors leads to reduction of Kir current in oligodendroglia within 1 s, most probably via a membrane-delimited pathway. Surprisingly, the G protein involved in this Kir inhibition is pertussis toxin-sensitive. Single channel recordings and the identification of mRNA indicate that the Kir involved is of the Kir2.0-type rather than the Kir3.0-type (Karschin, A. *et al.*, 1994; Karschin and Wischmeyer, 1995). Their conclusion is that the G protein inhibits the activity of Kir2.0-like channels in oligodendroglia via a membrane-delimited mechanism. [See Nilius et al. (1993) and Olesen and Bundgaard (1993) for the modulation of Kir channels in endothelial cells. For studies on Kir channels in endocrine cells, see Yamashita *et al.* (1988), Pennefather *et al.*, (1988), Einhorn *et al.*, (1991, 1993), and Takano *et al.* (1994).]

The transmitter-induced modulations are complicated and in several instances the conclusions appear to conflict with one another. Indeed, the modulations are complicated events, and the basic principle is not clearly understood. This state may have originated from several sources: (1) Some of the endogenous Kir channels may be different from the cloned Kir channels. Unknown splice variants may exist, and a small structural variation may result in large differences in the mode of modulation. (2) Experiments on heterologous systems are dependent on the signal transduction machinery endogenously available to the host cells. (3) Accessory proteins and cytoskeletal elements may influence the signal transductions.

6
Physiological Function of Strong Inward rectifiers.

Inward rectifiers of the Kir3.0 family are coupled to G-proteins and are activated by Gβγ. They play roles of great physiological importance in the regulation of electrical excitability through nervous action, roles in the central nervous system and in the heart, which have been discussed in detail in the previous two chapters. Modulation of channel P_{open} is a common means of inhibiting cells or of exciting them as through down regulation of constitutively active channels. Here we ask what the physiological significance of inward rectification might be and we shall concentrate on certain of the physiological roles of strong inward rectifiers, principally of the Kir2.0 family and principally in excitable cells.

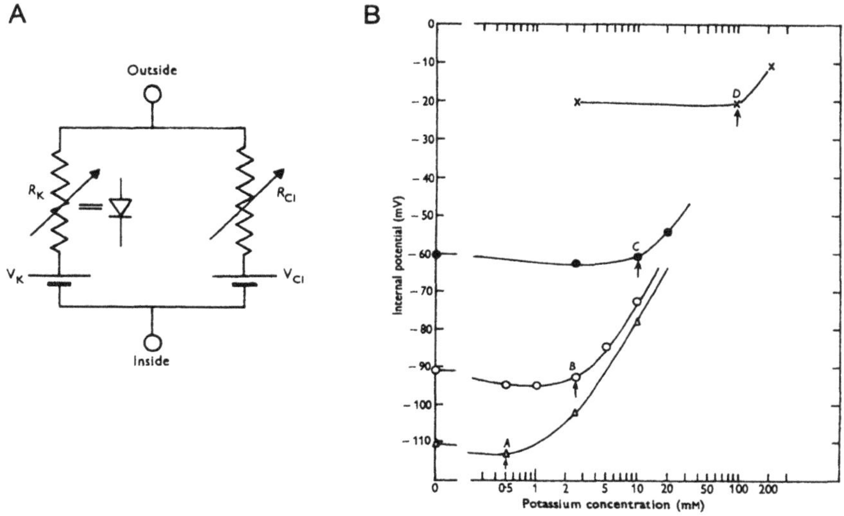

Fig. 14. Properties of inward rectification: activation and deactivation by changes in $[K^+]_o$.
A. Circuit diagram of the resting membrane of skeletal muscle. The resting potential is determined by K^+ and Cl^- conductances ($1/R_K$ and $1/R_{Cl}$), with current through these being driven batteries whose emfs (V_K and V_{Cl}) are set by the concentration gradients for K^+ and Cl^- respectively. Both conductance are variable, but that for K^+ is represented as a diode.
B. The instantaneous effects of changing $[K^+]_o$ at constant $[Cl^-]_o$. Single dissected muscle fibre of frog were equilibrated in solutions containing Cl^- at 120mM and K^+ at 0.5 (Δ), 2.5 (O), and 10mM (●). Fibres were also equilibrated in a 95mM K^+, 214mM Cl^- solution. Increasing $[K^+]_o$ from the value in the equilibration solution resulted in an instantaneous change in resting potential close to that expected from Nernst. In contrast, lowering $[K^+]_o$, which deactivated inward rectifier K^+ channels, results in little instantaneous change in resting potential, which was then determined principally by the unchanged E_{Cl}. Over time, the Donnan equilibrium would have been restored by exit of KCl from the muscle fibre.
Reproduced from Hodgkin & Horowicz (1959b), by permission.

In their initial description of the (V–E_K) dependence of inward rectifiers, Hodgkin & Horowicz (1959b) likened the potassium permeability mechanism in skeletal muscle to a diode, letting K^+ in but not out across the cell membrane (Fig. 14A). This is a useful electrical model, encapsulating an essential if surprising property of the permeability mechanism. But it can lead to an overstatement of how firmly K^+ channels are shut over the range of membrane potentials found under physiological conditions. For Kir2.1 channels, channel P_{open} is about 30% of its maximum at E_K and the membrane must be depolarised by several tens of mV to bring P_{open} to its minimum value.

The Kir2.1 channels are, then, of great importance in setting cellular resting potentials and in determining how easily membrane potential may be changed from rest. Their inwardly rectifying properties allow Kir channels to play an important role in conservation of the ionic gradients that exist across cell membranes, particularly helping cells conserve K^+. Their low conductance under depolarisation limits K^+ efflux and so also reduces the need for Na^+ or Ca^{2+} entry during action potentials. In this way, they help conserve the energy gradient stored as the Na^+ battery. Since Kir channels tend to close under depolarisation, depolarisation of cell membranes occurs initially against the brake of an active K^+ conductance that is gradually reduced and then removed. In Hille's description (Hille, 1992) the initial depolarisation is like the opening of a door on the latch, depolarisation being opposed by K^+ efflux which stabilises the membrane potential; sufficient depolarisation will open the latch and permit the depolarization to proceed with little opposition from K^+ efflux. Depolarisation may then have a self-reinforcing element, since it reduces K^+ permeability, which in turn makes depolarisation easier. Thus, as Kandel & Tauc (1966) pointed out when they discovered neuronal inward rectification, the property will aid the summing of EPSPs. The self-reinforcing element in depolarisation may also aid the spread of excitation along dendrites in neurones (Wessel *et al.*, 1999) and into the T-system in skeletal muscle and heart. In addition, in cochlear hair cells, which are electrically tuned by ionic conductance to the sound frequencies to which they best respond (Crawford & Fettiplace, 1980), Kir channels play a central role in such tuning (Goodman & Art, 1996).

Kir2.1 channels at least also appear to play key roles in development, probably through their role in setting high negative resting potentials, close to E_K. Kir2.1 expression hyperpolarises myocytes and aids fusion to form myoblasts during development of skeletal muscle (*e.g.* Fischer-Lougheed *et al.*, 2001). It may be added that *Kir2.1-/-* animals have a cleft secondary palate (Zaritsky *et al.*, 2000). Also, Andersen's syndrome, which includes developmental abnormality as well as defects of excitability of heart and skeletal muscle, is often (but not always) associated with mutation of Kir2.1 (Plaster *et al.*, 2001).

Since they tend to shut under depolarisation, inward rectifiers are important in determining the shape of action potentials. They both help maintain the plateau of long action potentials. Since they do carry outward currents, and perhaps quite large outward currents at certain times (Ishihara, 1997; Ishihara & Ehara, 1998), however, Kir channels also contribute significantly to action potential repolarisation. The activation of inward rectifiers by an increase in extracellular K^+, including the enhancement by K^+_o of K^+ efflux, is of importance in some cellular roles, particularly in permitting K^+ to act as a vasodilator of certain vascular beds. Roles for Kir2.1 in repolarisation of the action potential of ventricular muscle and in K^+ induced vasodilatation have now been firmly established by work using animals with Kir2.0 genes knocked out (Zaritsky *et al.*, 2000; Zaritsky *et al.*, 2001).

Where there are differences in $[K^+]_o$ over the surface of cells, channels may aid uptake of K^+ moving downhill *inwards* through inward rectifier channels. Channel P_{open} will be raised in such regions, while ionic concentrations and permeabilities over the cell as a whole will determine what the cellular resting potential will be. Thus the high conductance at negative voltage may also be used to help conserve cell K^+. Kir channels are believed to play a major role in K^+ homeostasis by glial cells in the nervous system and in the transverse tubular (T-) systems of skeletal and cardiac muscle. Channels also subserve K^+ cycling in basolateral membranes of transporting epithelia, (Kim *et al.*, 2000). This phenomenon occurs principally through Kir7.1, which pass large outward K^+ currents and whose P_{open} is largely independent of $[K^+]_o$; these channels are often co-localised with the Na^+-K^+ ATPase (Nakamura, *et al.*, 1999; Derst *et al.*, 2001a; Kusaka *et al.*, 2001).

6.1 Resting potential and the activation and de-activation of Kirs by K^+_o

6.1.1 Resting potential of skeletal muscle

Inward rectification was discovered in skeletal muscle (Katz, 1949; Hodgkin & Horowicz, 1959b). It is well established that, even if Cl^- permeability is higher, Kir activity is the principal long term determinant of the resting potential of skeletal muscle (Hodgkin & Horowicz, 1959b). However, the activation of K^+ channels by K^+_o and their deactivation by a reduction in $[K^+]_o$ results in unexpected changes in the relationship between resting potential and $[K^+]_o$, depending on other ionic permeabilities. In skeletal muscle fibres equilibrated in solutions containing different concentrations of KCl, suddenly lowering $[K^+]_o$, which shuts K^+ channels, results in the resting potential now being principally dependent on E_{Cl}; as a consequence, resting potential becomes largely independent of E_K, though only in the short term (Fig. 14B; Hodgkin & Horowicz, 1959b).

6.1.2 Expression of Kir channels – resting potential and development

In developing skeletal muscle, Kir2.1 expression appears to aid fusion of myoblasts, owing to the hyperpolarisation that occurs as Kir2.1 brings the resting potential close to E_K (Fischer-Lougheed *et al.*, 2001). This involvement was first demonstrated in chick muscle (Shin *et al.*, 1997), but was later shown to be the case with human muscle also (Liu *et al.*, 1998; Fischer-Lougheed *et al.*, 2001). Myoblast fusion to form myotubes depends on the entry of Ca^{2+} and follows expression of K^+ channels, first of the *eag* family

(Bernheim *et al.*, 1996; Bijlenga *et al.*, 1998, 2000) and then of Kirs. Fusion of human myoblasts was inhibited by Ba^{2+} or Cs^{+}, which prevent the hyperpolarisation caused by expression and activity of Kirs (Liu *et al.*, 1998). A similar pattern - expression first of *eag*-like, then of Kir channels that cause hyperpolarisation towards E_K - is seen during neuronal development from neural crest (in quail; Arcangeli *et al.*, 1997).

Transcripts for mRNA of both Kir2.1 and 2.2 were found in human myoblasts, with levels increasing during differentiation (Fischer-Lougheed *et al.*, 2001). Electrical recording suggested that Kir2.1 was the dominant species of channel in the surface. When antisense oligonucleotides were used to inhibit channel expression, inhibition of the expression of Kir2.1 inhibited fusion; inhibition of the expression of Kir2.2 did not. The K^{+} conductance hyperpolarises the myoblast sufficiently that T-type Ca^{2+} channels, that are also now expressed, become functional. This T-type channel function depends on the removal of their voltage dependent inactivation by the now negative resting potential.

While Kir2.1 clearly aids myoblast fusion, its expression may not to be an absolute requirement. Humans with Andersens' syndrome usually lack a functional Kir2.1 (Plaster *et al.*, 2001). Apart from symptoms of myotonia and periodic paralyses associated with the unstable resting potential of their skeletal muscle fibres, a muscle biopsy from a patient shows muscle fibres, but with tubular aggregates (Plaster *et al.*, 2001). The morphology of skeletal muscle in *Kir2.1-/-* mice has not yet been reported (Zaritsky *et al.*, 2000, 2001).

Mice that lack Kir2.1 show a cleft secondary palate and a narrowed lower mandible (Zaritsky *et al.*, 2000). Humans with Andersen's syndrome, lacking functional Kir2.1, have dysmorphic features (Plaster *et al.*, 2001). These features include short stature, curvatures of the spine (scoliosis) and of fingers or toes, and changes in the shape of the face (broad forehead, wide set eyes, small chin). Such errors of development imply that normal bony development may depend on Kir2.1 expression. Karschin & Karschin (1997) have shown that Kir2.1 is expressed before day E12 in embryonic rat and that this expression is primarily present in head, limbs and body, associated with bone structures. Plaster *et al.* (2001) argue for a role for Kir2.1 in craniofacial and skeletal morphogenesis. The exact cellular role of Kir2.1 is unclear, but is likely to be the subject of intense scrutiny in the immediate future. It is however known that Kir2.1 channels are expressed in osteoclasts, where it is possible they play a role in osteoclast formation through cellular fusion comparable to their role in myoblast formation. Among other syncytia, syncytiotrophoblasts also strongly express Kir2.1 (Clarson *et al.*, 2001).

These roles of Kir channels will undoubtedly be pursued further in the immediate future. The development of gene knockout animals will provide an important tool in enhancing understanding of these roles. For the present expression of Kir2.1 in developing skeletal muscle makes the membrane potential more negative by increasing K^{+} permeability and K^{+} efflux.

6.1.3 K^+ acts as a vasodilator by activating Kir channels and enhancing K^+ efflux

The activation of inward rectifier K^+ channels by extracellular K^+ is of particular importance in the vascular system. K^+ is an important vasodilator in cerebral, coronary and skeletal muscle vasculature (*e.g.* Kuschinsky *et al.*, 1972). Edwards *et al.* (1988a,b) were the first to show that while proximal arterioles arising from the middle cerebral artery of rat were depolarised if external K^+ was raised from 5 to 10mM, distal segments of arterioles were hyperpolarised. Kir channels were thought to generate the hyperpolarisation found in these resistance vessels. The resting potential was considerably positive to E_K, owing to permeability to other ions. Increasing $[K^+]_o$ opened Kir channels and increased K^+ efflux so that the membrane potential became more negative (Fig. 15A). Similar evidence that inward rectifier potassium channels underlie this response to elevations of K^+ have been obtained by Quayle *et al.* (1996) for coronary vessels and by Knot *et al.* (1996) for both coronary and cerebral blood vessels. In agreement with Edwards *et al.* (1988), Quayle *et al.* (1996) found higher densities of inward rectifier K^+ currents in myocytes from more distal arterioles of smaller diameter (Fig. 15B). The K^+ ions producing vasodilatation will be released from neurones or from muscle fibres as a result of the firing of action potentials so that vasodilatation can be matched to electrical activity. It has been argued that, in the central nervous system, astrocytes, which collect K^+ from active neurones (see below), deliver it to blood vessels through their end feet to produce vasodilatation more speedily (Paulson & Newman, 1987).

Perhaps more controversially, Edwards *et al.* (1998) have proposed that endotholium derived hyperpolarising factor (EDHF) is K^+ and that endothelial release of K^+ occurring through Ca^{2+} activated K^+ channels acts to relax smooth muscle partly through Kir. This proposal was supported by measurements of K^+ concentrations between endothelium and smooth muscle with ion sensitive electrodes and by the finding that low concentrations of Ba^{2+}, fairly specific for strong inward rectifiers, block the hyperpolarising response (Edwards *et al.*, 1998). Part of the response may also depend on activation of the Na^+-K^+ ATPase (*e.g.* McCarron & Halpern, 1990) since ouabain (albeit at 1mM) also reduces vasodilatation (Edwards *et al.*, 1998). Thus Edwards *et al.* (1998) propose a general role for K^+ as a vasodilator. Even in kidney, where its involvement is controversial (*e.g.* Prior *et al.*, 1998), there is evidence of K^+ activation of Kir being an important regulator in rat renal afferent arterioles at low perfusion pressures (Chilton & Loutzenhiser, 2001).

Bradley *et al.*, (1999) used RT-PCR to demonstrate expression of Kir2.1, but not of 2.2 or 2.3, in smooth muscle from rat basilar, coronary and mesenteric arteries. Chrissobolis *et al.* (2000) have also shown Kir2.1 in rat basilar artery by RT-PCR. Bradley *et al.* (1999) compared blockage by extracellular

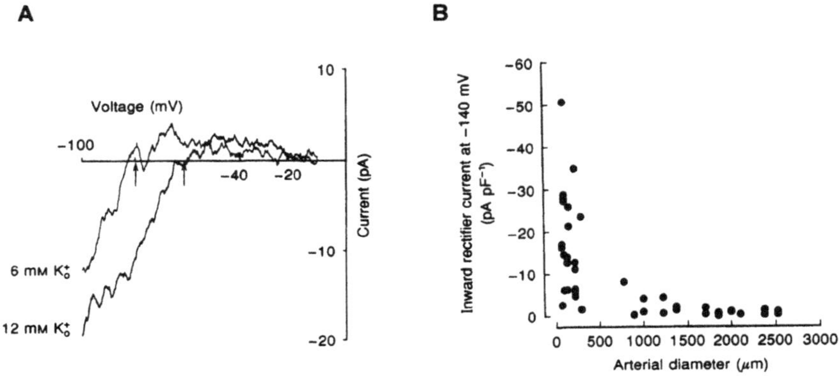

Fig. 15. Inward rectifier of smooth muscle from porcine coronary artery. **A.** Inward rectifier current voltage relations from smooth muscle cells isolated from porcine coronary artery at 2 external K^+ concentrations. The currents were dissected from other membrane currents using blockage by adding Ba^{2+}. At -40mV, for example, efflux of K^+ is greater with 12mM $[K^+]_o$ than with 6mM $[K^+]_o$, the two current voltage relations crossing over each other at about -50mV membrane potential.Thus adding K^+ activated inward rectifier K^+ channels, increasing K^+ efflux, and would hyperpolarise the smooth muscle cells in physiological conditions. **B.** If current density is measured in smooth muscle cells from arteries of different diameter, there is a trend to higher densities in smaller, resistance vessels. Abscissa, arterial diameter (µm); ordinate Kir current density (pA.pF). Reproduced from Quayle *et al.* (1998), by permission.

Mg^{2+}, Ca^{2+}, Ba^{2+} and Cs^+ of Kir channels in native arterial smooth muscle with that of Kir2.1 expressed in oocytes and found the characteristics of blockage to be essentially the same. They concluded that Kir2.1 was the route through which vasodilatation was produced (Bradley *et al.*, 1999). Chrissobolis *et al.* (2000) tested *in vivo* the effects of changing the $[K^+]$ of cerebrospinal fluid on diameter of basilar artery. They observed vasodilatation when $[K^+]$ was increased from 3 to up to 30mM. Ba^{2+} at 30 or 100 µM blocked this dilatation, but ouabain at 1 - 100 µM did not. Inhibition of nitric oxide synthase also did not interfere with the vasodilator effect of K^+, indicating that the effect is independent of the vasodilator action of nitric oxide. Both groups (Bradley *et al.*, 1999; Chrissobolis *et al.*, 2000) thus argue that Kir2.1 is the route through which K^+ relaxes blood vessels since these channels are expressed in vascular smooth muscle, and since the physiological and pharmacological properties of the channels found electrophysiologically resemble those of Kir2.1.

The involvement of Kir2.1 now seems certain after measurements in cerebral blood vessels from neonatal mice that have deletions of either the *Kir2.1* or the *Kir2.2* genes (Zaritsky *et al.*, 2000). *Kir2.2-/-* mice show a normal response to K^+. But *Kir2.1-/-* mice lack any vasodilator response to increasing $[K^+]_o$. This is true whether blood vessels are constricted by pharmacological means using a thromboxane A_2 mimetic or through a raised, 40 or 80mm Hg intravascular pressure (Zaritsky *et al.* 2000). In wild type animals of the same age (P1), vasodilatation is induced in the same experimental conditions and this dilatation is Ba^{2+}-sensitive. Thus K^+ activation of K^+ efflux - permitted by the gating of Kir2.1 by intracellular polyamines and by K^+_o as described in Section 3 - is a powerful physiological vasodilator mechanism.

Kir2.1 may play roles in activation of endothelial cells also. Inward rectifier K^+ channels were known to be present in such cells, where they are inhibited by shear stress (Olesen *et al.*, 1988). Kir2.1 has been identified as the channel in bovine aortic endothelial cells (Forsyth *et al.*, 1997). In these cells, mRNA levels for Kir2.1 are reduced if endothelial cells are exposed to 30dyn.cm^{-2} shear stress for 24h.

6.2 Kir Channels Contribute to Action Potential Shape and Duration: Kir2.0 of Heart Muscle

Channels carrying the inward rectifier current I_{K1} set the resting potential of cardiac myocytes. Since channels shut under depolarisation, the plateau of the action potential is sustained through a reduction of the membrane conductance to a level lower than at rest (Weidmann, 1951). Outward currents flow during the initial depolarisation of the action potential and also during the final phase of repolarisation, as Ibarra *et al.* (1991) demonstrated by measuring Cs^+ sensitive currents during an action potential waveform applied under voltage clamp. In their experiments, outward currents were smaller during repolarisation than during the initial depolarisation, presumably because of increased occupancy of the channel by spermine, rather than Mg^{2+} or putrescine during long depolarisations (Ishihara, 1997; Ishihara & Ehara, 1998; see Fig. 9).

Although Kir2.1, 2.2, and 2.3 are expressed in heart, there is good evidence that Kir2.1 is the principal contributor to I_{K1}, together with a subsidiary role for Kir2.2. The evidence comes from *in situ* hybridisation studies (Brahmajothi *et al.*, 1996), from use of anti-sense technology (Nakamura *et al.*, 1998), and from gene knockout (Zaritsky *et al.*, 2001). *In situ* hybridisation studies showed expression of a member of the Kir2.0 subfamily in heart whose distribution mirrored that of the density of I_{K1}. Thus expression was universal in ventricular myocytes, common in atrium, but nearly absent from sinu-atrial node (<7% of cells), consistent with the virtual absence of the current in the pacemaker (Brahmajothi *et al.*, 1996). Antisense olignucleotides designed specifically against Kir2.1, and with no inhibitory action on expression of Kir2.2 or 2.3, were shown to inhibit I_{K1} in rat ventricular myocytes but to have no effect on L-type Ca^{2+} current or on transient outward K^+ current (Nakamura *et al.*, 1998).

Knockouts of Kir2.1 and 2.2 both have effects on inward rectifier K^+ currents of heart, as measured in neonatal animals. These currents were abolished in *Kir2.1-/-*. In these animals, however, Kir2.3 expression in heart was enhanced, though without discernible physiological effect. While currents were abolished in *Kir2.1-/-*, they were also reduced by about 50% in *Kir2.2-/-*, implying that this channel subunit contributes to I_{K1}. This finding is somewhat surprising in the light of evidence against heteromeric assembly of Kir2.0 family members (Tinker *et al.*, 1996; see Section 2); but the possibility exists that some channels contributing to I_{K1} are heteromers. Ca^{2+} currents and those carried by Kv are unaltered in either *Kir2.1-/-* or *Kir2.2-/-*.

The result of abolition of I_{K1} in *Kir2.1-/-* was an increase in spontaneous activity in dissociated ventricular myocytes and a substantial increase in action potential duration, with decay time doubled (Fig. 16). A related effect – long QT intervals in the ECG – is seen in humans with Andersen's syndrome, caused by mutations of Kir2.1 that generate a dominant negative, non-conducting form of the channel (Plaster *et al.*, 2001). The ECG of *Kir2.1-/-* also shows long QT intervals (Zaritsky *et al.*, 2001), that of *Kir2.2-/-* does not. Assessment of the current lost in the Kir2.1 channel knockout suggests that this K^+ channel contributes to all phases of repolarisation of the action potential (Zaritsky *et al.*, 2001). This assessment is surprising given the belief that Kir contributes only to late, phase 3 repolarisation (Ibarra *et al.*, 1991). Zaritsky *et al.* (2001) consider the possibility that a change in the action potential upstroke may alter activation or inactivation of transient outward current or that Kirs contribute more outward current than had hitherto been supposed.

Although the resting potential of myocytes will be less well clamped close to E_K so that isolated myocytes show increased spontaneous activity, neither *Kir2.1-/-* nor *Kir2.2-/-* animals show signs of cardiac arrhythmia. *Kir2.1-/-* animals have bradycardia, perhaps owing to changes in the peripheral vasculature, with loss of a mechanism of vasodilatation (Zaritsky *et al.*, 2000; see above) or to changes in autonomic function. This bradycardia in *Kir2.1-/-* is the principal cause of the long QT interval. In Andersen's syndrome, abnormalities of cardiac electrical excitability include presence of ectopic foci in the ventricle, ventricular tachycardia, *torsades de pointes*, and cardiac arrest (Plaster *et al.*, 2001), symptoms apparently more severe than described in *Kir2.1-/-* mice. Humans with Andersen's syndrome due to Kir2.1 mutation however live to adulthood, while *Kir2.1-/-* mice live < 1 day. Thus the comparison may be of different stages of cardiovascular development.

Changes in the excitability of skeletal muscle also occur in Andersen's syndrome because of the central role of Kir2.1 in setting resting potential. In members of different families with the disorder, paralysis of muscle may be associated with hypo-, hyper- or normokalaemia (Plaster *et al.*, 2001).

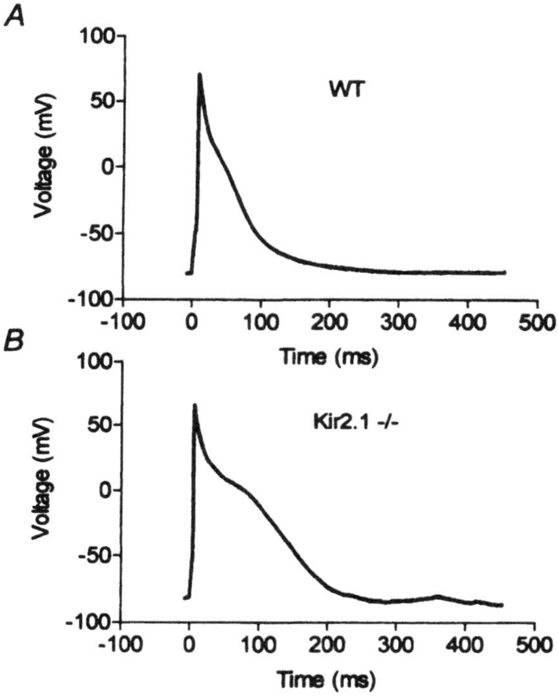

Fig. 16. Action potentials of ventricular myocytes in wild type and in Kir2.1$^{-/-}$ mice. Action potentials were recorded from ventricular myocytes of wild type (A) and Kir2.1$^{-/-}$ mice – the lack of inward rectifier Kir21, channels resulted in prolongation of the action potential. Surprisingly, all phases of repolarisation were slowed in the absence of Kir2.1. Reproduced from Zaritsky et al. (2001) by permission.

6.3
K^+ homeostasis - K^+ Activation of Kir Channels and K^+ Influx

Several authors have suggested a role for inward rectifiers in helping to dispose of elevated external K^+. Glia are able to 'siphon' K^+ from regions of high to regions of low extracellular K^+ concentration (Orkand *et al.*, 1966). Involvement of Kirs has been proposed at least for glial uptake of K^+ (Newman, 1984; Brew *et al.*, 1986). But it is also likely to be an important mechanism for K^+ homeostasis in the transverse tubular system of heart and skeletal muscle (Wallinga *et al.*, 1999).

6.3.1 Requirements for entry of K^+ through Kir.

K^+ absorption through Kir requires that E_K must be locally elevated above the resting potential of the cell – *i.e.* that $(V - E_K) < 0$ – so that K^+ may enter cells downhill. There are at least three mechanisms by which this transfer can occur. These are first, electrogenic activity of the Na^+-K^+ ATPase; secondly, disturbance of the Donnan equilibrium (Hodgkin & Horowicz, 1959b); and thirdly, a non-uniform distribution of ionic concentration (Hodgkin & Horowicz, 1960).

Membrane potential will become negative to E_K during periods when the Na^+-K^+ ATPase is strongly activated. Tetanic stimulation of a neuron results in an increase of $[Na^+]_i$, which in turn will cause hyperpolarisation (Kerkut & Thomas, 1965; Nakajima & Takahashi, 1966; Rang & Ritchie, 1968; Baylor & Nicholls, 1969; Thomas, 1972; Rakowski *et al.*, 1989) owing to its electrogenic properties (efflux of $3Na^+$ coupled to influx of $2K^+$).

How large is the magnitude of $(V - E_K)$ during hyperpolarisation by the pump? When the Na^+-K^+ ATPase operates vigorously to extrude Na^+ from Na^+-loaded skeletal muscle fibres, the membrane potential may become as much as 20mV negative to E_K (Adrian & Slayman, 1966). Simultaneous measurement of V and of $[K^+]_i$ or of V and $[K^+]_o$ were performed in frog motoneurons (Grafe *et al.*, 1982). Estimates from these published results suggest that V may be as much as 8mV negative to E_K after tetanic stimulation. Such effects may be expected to be greater in smaller neurons, which have a larger surface/volume ratio. Nevertheless, the presence of Kir channels will also limit the amplitude of the hyperpolarisation, which will raise K^+ permeability. In turn, the presence of Kir channels will mean that it is almost exclusively K^+ that will be absorbed during small hyperpolarisations beyond E_K.

The presence of high Cl^- permeability in muscle, for example, means that a change of Donnan equilibrium can result in K^+ uptake. When $[K^+]_o$ is suddenly increased experimentally (*e.g.* by replacing Na^+) at constant $[Cl^-]_o$ in isolated single skeletal muscle fibres, membrane potential does not immediately change to its new value, but takes time (tens of min) fully to attain its depolarised, equilibrium value. Immediately after the increase in $[K^+]_o$, E_{Cl} is unchanged and the membrane potential takes a value somewhere between the new E_K and the unchanged E_{Cl} (though it will be closer to E_K, as Fig. 14B shows). The Donnan equilbrium will subsequently be slowly re-established. During most of this period, V is at least slightly negative to E_K and K^+ entry occurs through Kir channels, which are present and open both in the surface and in transverse tubular membranes. This K^+ entry will be aided by the activation of Kir channels by the elevated $[K^+]_o$.

Local changes in $[K^+]_o$ are also known to occur in the nervous system, particularly in regions of high electrical activity or around regions of synaptic contact (see *e.g.* Horio & Kurachi, 1999). These increases in $[K^+]_o$ must be kept under homeostatic control and this control is exerted by siphoning of K^+ (Orkand *et al.*, 1966) or by uptake of K^+ by glia (for review, see Horio & Kurachi, 1999). Non-uniform, local changes in K^+ concentration can also occur quite easily in muscle, for example, owing to the limited volume of the lumen of the transverse tubular system. Hodgkin & Horowicz (1960) made rapid changes of $[K^+]_o$ around dissected single muscle fibres and showed that inward rectifier K^+ channels were present partly in the membrane lining a restricted element of the extracellular space. Nakajima *et al.* (1973) showed this restricted space was the lumen of the transverse tubular or T-system, through which action potentials propagate to stimulate excitation-contraction coupling. In the experiments of Hodgkin & Horowicz (1960), if $[K^+]_o$ was elevated very briefly (for a second or less), its subsequent reduction led to rapid repolarisation. But if $[K^+]_o$ was raised for a few s, repolarisation was delayed owing to the rise in $[K^+]$ in the tubular lumen. K^+ had now had time to diffuse into the T-system and took time to diffuse out again. In these conditions, where tubular $[K^+]$ ($[K^+]_T$) is higher than bulk $[K^+]_o$, resting potential will be a compromise between the different E_Ks at the surface and across the T-system membrane, and E_{Cl}. The high chloride permeability of skeletal muscle will help keep the resting potential negative. Within the T-system, however, E_K will be positive to resting potential, with the result that K^+ will be able to move in across the tubular membrane.

6.3.2
K^+ Homeostasis in the Nervous Systems

Kirs play a major part in glial regulation of $[K^+]_o$ in the central nervous system and probably in peripheral nerve (Horio & Kurachi, 1999). Such absorption of K^+ has been proposed as the mechanism for glial uptake, which has been intensively studied in retinal Müller cells (Newman, 1984; 1993; Brew *et al.*, 1986). Much of this buffering occurs through channels of the Kir4.0 subfamily (*e.g.* Ishii *et al*, 1997), which, being weak rectifiers and permitting substantial K^+ efflux, are well suited to a role in uptake of K^+ in regions of high concentration and its release in regions of low concentration. Kir 4.0 channels are found in ependymal cells in salmon brain (Kubo *et al.*, 1996). They are also found in oligodendrocytes (Takumi *et al.*, 1995), where they may be modulated by the action of certain transmitters (Karschin, A. *et al.*, 1994; Horio & Kurachi, 1999). But Kir2.0 members are present in Schwann cells (Mi *et al.*, 1996) and Kir2.3 are present in astrocytes, particularly in type I astrocytes (Perillán *et al.*, 2000). Other members of the subfamily also appear to be expressed in certain glial cells (Stonehouse *et al.*, 1999; Leonoudakis *et al.*, 2001). Kir channels may be distributed primarily to the fine processes of both type 1 and type 2 astrocytes (Barres *et al.*, 1990), permitting take up of K^+ principally into these processes (Horio & Kurachi, 1999).

A member (or members) of the Kir2.0 subfamily is (are) present in the microvilli of Schwann cells, in the region immediately around the node of Ranvier (Mi *et al.*, 1996). The suggestion is that they play a role in uptake of K^+ released from nodes during action potential propagation, which strong rectifiers will readily be able to do. Though their molecular identity was not then known, the density of Kir channels is greater in Schwann cells of myelinating axons (Wilson & Chiu, 1990), current density falling by 80-90% as the number of lamellae rose from 6 to 21. This reduction was proposed as reflecting the lack of a need to buffer K^+ in internodal regions after myelin had been laid down.

6.3.3 Kir and Actions Potential in Skeletal Muscle

During activation of skeletal muscle, an action potential propagates through the disc of the T-system network (*e.g.* Costantin, 1970; Adrian & Peachey, 1973; Bastian & Nakajima, 1974). Many lines of evidence point to a T-tubular location for Kirs. Thus, hyperpolarisation, imposed under current or voltage clamp conditions, reduces tubular $[K^+]_o$ (Adrian & Freygang, 1962a; Adrian *et al.*, 1970; Standen & Stanfield, 1979), since K^+ moves in across the T-system membrane more quickly than it is replenished from the bulk extracellular fluid. Measurements of T-tubular membrane potential with potentiometric dyes under appropriate conditions also indicated that Kir channels are present at approximately equal density on surface and in T-tubular membranes (Heiny *et al.*, 1983). Work on heart muscle (Leonoudakis *et al.*, 2001) has shown that Kir2.2 (and probably 2.1) is tethered in the T-system membrane by its association with the scaffolding protein SAP97.

Action potential propagation will increase tubular $[K^+]$, since the lumen of the T-system is small compared with the fibre volume (about 0.003 of the total fibre volume in frog sartorius; *e.g.* Adrian *et al.*, 1970). The net rise in concentration after a single action potential is thought to be in the order of 0.4mM (Kirsch *et al.*, 1977), and this will be greater for a train of action potentials. This rise in concentration will activate Kir channels, since their P_{open} is dependent on $[K^+]_o$. Such activation will have two effects. First during the repolarisation of the action potential, it will enhance K^+ efflux through Kir, helping to accelerate action potential repolarisation. Leonoudakis *et al.* (2001) have proposed this role for Kir2.2 in heart muscle. Secondly, once repolarisation is complete, the membrane potential may fall below T-system E_K, since $[K^+]_T > [K^+]_o$, and permit K^+ to return downhill to sarcoplasm.

These hypotheses have yet to be tested experimentally. However, Wallinga *et al.* (1999) developed a model of action potential propagation in skeletal muscle from those used by Adrian & Peachey (1973) and others. Cable theory was applied to the T-system network to calculate the propagating action potential, the currents that produce it and the changes of $[K^+]_T$. Assumptions were about the relative density (h) of Kir and Kv channels and of the Na^+-K^+ ATPase in T-system and on the surface. For Kir, h = 1 following Heiny *et al.* (1983), while for Kv, it was 0.45 and for the Na^+-K^+ ATPase 0.1. In these conditions, Kir channels contributed currents nearly $^1/_3$ the amplitude of those from Kv during action potential repolarisation in the innermost part of the T-system.

In the computation of Wallinga *et al.* (1999), the surface and tubular action potential became smaller and broader during trains of action potentials at 40Hz. But this effect was limited by K^+ clearance catching up with accumulation during the train owing to the return of K^+ to sarcoplasm by Kirs. Action potential reduction was much more extreme (about 50mV rather than 10mV by the 40th action potential) if the model lacked inward rectifier K^+ channels in the T-system. This reduction was due to the increased depolarisation that results from K^+ clearance now being slower than accumulation and the consequent inactivation of voltage gated sodium channels. In this condition, excitation-contraction coupling was compromised.

Thus inward rectifiers play a significant role in K^+ homeostasis in excitable and other cells.

6.4 Concluding Comment

In an autobiography (Hodgkin, 1992), Alan Hodgkin described almost every aspect of his work both in physiological science and in the development of airborne radar during the 1939-45 war. One contribution only is omitted. This is any description of his work over 40 years ago with Paul Horowicz of the essential properties of inward rectifier potassium channels (Hodgkin & Horowicz, 1959b; 1960). One can only speculate about the reasons for the omission - that it may have arisen because the permeability mechanism seemed odd, anomalous and surprising. As they put it (Hodgkin & Horowicz, 1959b) at the time 'neither the physiological significance nor the physical nature of this rectification (was) understood. The matter is particularly puzzling . . . '. The great diversity of inward rectifier K^+ channels shown by molecular biology has emphasised just how important these channels are. Much is now known about the physical nature of inward rectification. Their properties fit them for several physiological roles, a few of which we have described.

Acknowledgement: We thank Mordecai Blaustein for reading the manuscript and giving us valuable suggestions. Supported by NIH grants AG06093 and MH57837 and by the Wellcome Trust and Medical Research Council.

Abbreviations

AA:	arachidonic acid
ABC:	ATP binding cassette
AKAP:	the A kinase anchoring protein
CIPP:	channel-interacting PDZ domain protein
C-TAG:	C-terminal association with GLGF
EDHF:	endotholium derived hyperpolarising factor
GAP:	GTPase-activating protein
GGL domain:	G protein gamma subunit-like domain
GPCR:	G protein-coupled receptor
GST:	glutathione-S-transferase
Gt:	transducin
IP_3:	inositol trisphosphate
K_{ACh}:	ACh-activated Kir channel
Kir channels:	Inward rectifier K^+ channel
KirNB channels:	Kir channels in nucleus basalis neurons
Kv:	voltage-gated K^+ channel
LC:	locus coeruleus
MAGUK:	membrane associated guanylate kinase
MEL:	murine erythroleukaemia
MTSEA:	methanthiosulphonate ethylammonium
MTSET:	methanthiosulphonate ethyltrimethylammonium
ODC:	ornithine decarboxylase
PIP_2:	phosphatidylinositol 4,5-bisphosphate
PIP_3:	phosphatidylinositol 3,4,5-trisphosphate
PKA:	cyclic AMP-dependent protein kinase
PKC:	protein kinase C
PLC:	phospholipase
PP2A:	phosphoprotein phosphatase 2A
QA:	quaternary alkylammonium
RGS:	Regulators of G protein signalling.
SAMDC:	S-adenosyl methionine decarboxylase
SUR:	sulphonylurea receptor
TBA:	tetrabutylammonium
TEA:	tetraethylammonium
TRH:	thyrotropin-releasing hormone

References

Abrams CJ, Davies NW, Shelton PA, Stanfield PR (1996). The role of a single aspartate residue in ionic selectivity and block of a murine inward rectifier K^+ channel Kir2.1. Journal of Physiology 493: 643–649.

Adams PR, Brown DA, Constanti A (1982). Pharmacological inhibition of the M-current. Journal of Physiology 332:223–262

Adrian RH (1964) The rubidium and potassium permeability of frog muscle membrane. J. Physiol. 175: 134–159.

Adrian RH (1969). Rectification in muscle membrane. Progress in Biophysics 19: 341–369.

Adrian RH, Chandler WK, Hodgkin AL (1970). Slow changes in potassium permeability in skeletal muscle. Journal of Physiology 208: 645–668.

Adrian RH, Freygang W (1962a). The potassium and chloride conductance of frog muscle membrane. Journal of Physiology 163: 61–103.

Adrian RH, Freygang W (1962b). Potassium conductance of frog muscle under controlled voltage. Journal of Physiology 163: 104–114.

Adrian RH, Peachey LD (1973). Reconstruction of the action potential of frog sartorius muscle. Journal of Physiology 235: 103–131.

Adrian RH, Slayman CL (1966). Membrane potential and conductance during transport of sodium, potassium and rubidium in frog muscle. Journal of Physiology 184: 970–1014

Aghajanian GK, Wang YY (1986). Pertussis toxin blocks the outward currents evoked by opiate and α_2-agonists in locus coeruleus neurons. Brain Research 371:390–394

Aguilar-Bryan L, Nichols CG, Wechsler SW, Clement IV JP, Boyd III AE, Gonzalez G, Herrera-Sosa H, Nguy K, Bryan J, Nelson DA (1995). Cloning of the beta cell high-affinity sulfonylurea receptor: a regulator of insulin secretion. Science 268: 423–426.

Alagem N, Dvir M, Reuveny E (2001). Mechanism of Ba^{2+} block of a mouse inwardly rectifying K^+ channel: differential contribution by two discrete residues. Journal of Physiology 534: 381–393.

Albsoul-Younes AM, Sternweis PM, Zhao P, Nakata H, Nakajima S, Nakajima Y, and Kozasa T (2001) Interaction sites of the G protein β subunit with brain G protein-coupled inward rectifier K^+channel. Journal of Biological Chemistry, 276: 12712–12717.

Aleksandrov A, Velimirovic B, Clapham DE (1996). Inward rectification of the IRK1 K^+ channel reconstructed in lipid bilayers. Biophysical Journal 70: 2680–2687.

Almers W, Stanfield PR, Stühmer W (1983). Lateral distribution of sodium and potassium channels in frog skeletal muscle: measurements with a patch-clamp technique. Journal of Physiology 336:261–284

Andrade R, Malenka RC, Nicoll RA (1986). A G protein couples serotonin and $GABA_B$ receptors to the same channels in hippocampus. Science 234:1261–1265

Araneda RC, Lan J-Y, Zheng X, Zukin RS, Bennett MVL (1999). Spermine and arcaine block and permeate N-methyl-D-aspartate receptor channels. Biophysical Journal 76: 2899–2911.

Arcangeli A, Bianchi L, Becchetti A, Faravelli L, Coronnello M, Mini E, Olivotto M, Wanke E (1995). A novel inward-rectifying K^+ current with a cell-cycle dependence governs the resting potential of mammalian neuroblastoma cells. Journal of Physiology 489: 455–471.

Arcangeli A, Rosati B. Cherubini A, Crociani O, Fontana L Ziller C, Wanke E, Olivotto M (1997). HERG- and IRK-like inward rectifier currents are sequentially expressed

during neuronal development of neural crest cells and their derivatives. European Journal of Neuroscience 9: 2596–2604.
Armstrong CM (1966). Time course of TEA^+ induced anomalous rectification in squid giant axons. Journal of General Physiology 50: 491–503.
Armstrong CM (1969). Inactivation of potassium conductance and related phenomena caused by quaternary ammonium ion injected into squid axons. Journal of General Physiology 54: 553–575.
Armstrong CM (1971). Interaction of tetraethylammonium derivatives with the potassium channels of giant axons. Journal of General Physiology 58: 413-437.
Ashcroft FM, Stanfield PR (1983). The influence of the permeant ions thallous and potassium on inward rectification in frog skeletal muscle. Journal of Physiology 343: 407–428.
Axelrod D (1983). Lateral motion of membrane proteins and biological function. Journal of Membrane Biology 75:1–10
Baines RA, Uhler JP, Thompson A, Sweeney ST, Bate M (2001). Altered electrical properties in *Drosophila* neurons developing without synaptic transmission. Journal of Neuroscience 21: 1523–1531.
Bannister JPA, Young BA, Sivaprasadarao A, Wray D (1999). Conserved extracellular cysteine residues in the inwardly rectifying potassium channel Kir2.3 are required for function but not expression in the membrane. FEBS Letters 458: 393–399.
Barres BA, Chun LLY, Corey DP (1990). Ion channels in vertebrate glia. Annual Reviews of Neuroscience 13:441–474
Bastian J, Nakajima S (1974) Action potential in the transverse tubules and its role in the activation of skeletal muscle. Journal of General Physiology 63: 257–278.
Baukrowitz T, Schulte U, Oliver D, Herlitze S, Krauter T, Tucker SJ, Ruppersberg JP, Fakler B (1998). PIP_2 and PIP as determinants for ATP inhibition of K(ATP) channels. Science 282: 1141–1144.
Baylor DA, Nicholls JG (1969) After-effects of nerve impulses on signalling in the central nervous system of the leech. Journal of Physiology 203: 571–589
Beirão PSL, Davies NW, Stanfield PR (1994). Inactivating 'ball' peptide from Shaker B blocks Ca^{2+}-activated but not ATP-dependent K^+ channels of rat skeletal muscle. Journal of Physiology 474: 269–274.
Benson JA, Levitan IB (1983). Serotonin increases an anomalously rectifying K^+ current in the *Aplysia* neuron R15. Proceedings of the National Academy of Sciences USA 80:3522–3525
Berman DM, Wilkie TM, Gilman AG (1996). GAIP and RGS4 are GTPase-activating proteins for the G_i subfamily of G protein a subunits. Cell 86:445–452
Bernheim L, Liu J-H, Hamann M, Haenggeli CA, Fischer-Lougheed J, Bader CR (1996). Contribution of a non-inactivating potassium current to the resting potential of fusion-competent human myoblasts. Journal of Physiology 493: 129–141.
Bezanilla F, Armstrong CM (1972). Negative conductance caused by entry of sodium and cesium ions into potassium channels of squid axons. Journal of General Physiology 60:588–608.
Bianchi L, Roy M-L, Taglialatela M, Lundgren DW, Brwon AM, Ficker E (1996). Regulation by spermine of native inward rectifier K^+ channels in RBL-1 cells. Journal of Biological Chemistry 271: 6114–6121.
Bijlenga Occhiodoro Liu Bader Bernheim Fischer-Lougheed J (2000). T-type alpha 1H Ca^{2+} channels are involved in Ca^{2+} signaling during terminal differentiation (fusion) of human myoblasts. Proceedings of the National Academy of Sciences USA 97: 7627–7632.

Bijlenga PT, Liu J-H, Hamann M, Haenggeli CA, Fischer-Lougheed J, Bader CR (1998). An *ether-a-go-go* K^+ current $I_{h\text{-}eag}$ contributes to the hyperpolarization of fusion-competent human myoblasts. Journal of Physiology 512: 317–323.

Boim MA, Ho K, Shuck ME, Bienkowski MJ, Block JH, Slightom JL, Yang Y, Brenner BM, Hebert SC (1995) ROMK inwardly rectifying ATP-sensitive K^+ channel. II. Cloning and distribution of alternative forms. American Journal of Physiology 268: F1132–F1140.

Bond CT, Ammala C, Ashfield R, Blair TA, Gribble F, Khan RN, Lee K, Proks P, Rowe ICM, Sakura H, Ashford MJ, Adelman JP, Ashcroft FM (1995). Cloning and expression of the cDNA encoding an inwardly-rectifying potassium channel expressed in pancreatic beta cells and in the brain. FEBS Letters 367: 61–66.

Bond CT, Pessia M, Xia X-M, Lagrutta AM Kavanaugh MP, Adelman JP (1994). Cloning and expression of a family of inward rectifier potassium channels. Receptors and Channels 2: 183–191.

Bowie D, Mayer ML (1995). Inward rectification of both AMPA and kainate subtype glutamate receptors generated by polyamine-mediated ion channel block. Neuron 15: 453–462.

Bradley KK, Jaggar JH, Bonev AD, Heppner TJ, Flynn ERM, Nelson MT & Horowitz B (1999). Kir2.1 encodes the inward rectifier potassium channel in rat arterial smooth muscle cells. Journal of Physiology 515: 639–651.

Brahmajothi MV, Morales MJ, Liu S, Rasmusson RL, Campbell DL, Strauss HC (1996). In situ hybridization reveals extensive diversity of K^+ channel mRNA in isolated ferret cardiac myocytes. Circulation Research 78: 1083–1089.

Braun AP, Fedida D and Giles WR (1992) Activation of α1-adrenoceptors modulates the inwardly rectifying potassium currents of mammalian atrial myocytes. Pflügers Arch, 421:431–439.

Brazier SP, Ramesh B, Haris PI, Lee DC, Srai SKS (1998). Secondary structure analysis of the putative membrane associated domains of the inward rectifier K^+ channel ROMK1. Biochemical Journal 335:375–380.

Breitwieser GE, Szabo G (1985) Uncoupling of cardiac muscarinic and β-adrenergic receptors from ion channels by a guanine nucleotide analogue. Nature 317:538–540

Breitwieser GE, Szabo G (1988) Mechanism of muscarinic receptor-induced K^+ channel activation as revealed by hydrolysis-resistant GTP analogues. Journal of General Physiology 91: 469–493.

Brew H, Gray PTA, Mobbs P, Attwell D (1986). Endfeet of retinal glial cells have higher densities of ion channels that mediate K^+ buffering. Nature 324: 466–468.

Brown DA (1990) G-proteins and potassium currents in neurons. Annual Review of Physiology 52:215–242

Bünemann M, Hosey MM (1998) Regulators of G Protein signaling (RGS) proteins constitutively activate Gβγ-gated potassium channels. Journal of Biological Chemistry 273:31186-31190

Burgen ASV, Terroux KG (1953) On the negative inotropic effect in the cat's auricle. Journal of Physiology 120:449-464

Cabrera, J.L., De Freitas, F., Satpaev, D.K., and Slepak, V.Z. (1998) Identification of the Gβ5-RGS7 protein complex in the retina. Biochemical and Biophysical Research Communications **249**: 898–902.

Carlson KE, Woolkalis MJ, Newhouse MG, Manning DR (1986) Fractionation of the b subunit common to guanine nucleotide-binding regulatory proteins with the cytoskeleton. Molecular Pharmacology 30:463–468

Carmeliet E (1979). Voltage dependent block of inward going rectification in cardiac Purkinje fibers by external Cs^+ ions. Archives Internationales de Pharmacodynamie et de Thérapie 242: 296–297.

Casey PJ (1994) Lipid modifications of G proteins. Current Opinion in Cell Biology 6:219–225

Chan KW, Sui JL, Vivaudou M, Logothetis DE (1996) Control of channel activity through a unique amino acid residue of a G protein-gated inwardly rectifying K^+ channel subunit. Proceedings of the National Academy USA 93:14193–14198

Chan KW, Sui JL, Vivaudou M, Logothetis DE (1997) Specific regions of heteromeric subunits involved in enhancement of G protein-gated K^+ channel activity. Journal of Biological chemistry 272: 6548–6555

Chandy KG, Gutman GA (1993) Nomenclature for mammalian potassium channel genes. Trends in Pharmacological Sciences 14: 434.

Chilton L, Loutzenhiser R (2001). Functional evidence for an inward rectifier potassium current in rat renal afferent arterioles. Circulation Research 88: 152–158.

Cho H, Nam G-B, Lee, SH, Earm YE, Ho W-K (2001) Phosphatidylinositol 4,5-bisphosphate is acting as a signal molecule in α_1-adrenergic pathway via the modulation of acetylcholine-activated K^+ channels in mouse atrial myocytes. Journal of Biological chemistry 276: 159–164.

More Cho HC, Tsushima et al. to here from p 157

Choe H, Palmer LG, Sackin H (1999). Structural determinants of gating in inwardly-rectifying K^+ channels. Biophysical Journal 76: 1988–2003.

Choe H, Sackin H, Palmers LG (1998). Permeation and gating of an inwardly rectifying potassium channel: evidence for a variable energy well. Journal of General Physiology 112: 433–446.

Choi KL, Aldrich RW, Yellen G (1991). Tetraethylammonium blockade distinguishes two inactivation mechanisms in voltage-activated K^+ channels. Proceedings of the National Academy of Sciences USA 88: 5092–5095.

Chrissobolis S, Ziogas J, Chu Y, Faraci FM, Sobey CG (2000). Role of inwardly rectifying K^+ channels in K^+-induced cerebral vasodilatation *in vivo*. American Journal of Physiology 279: H2704–2712.

Chuang HH, Jan YN, Jan LY (1997). Regulation of IRK3 inward rectifier K^+ channel by m1 acetylcholine receptor and intracellular magnesium. Cell 89: 1121–1132.

Chuang HH, Prescott ED, Kong H, Shields S, Jordt S-E, Basbaum AI, Chao MV, Julius D (2001). Bradykinin and nerve growth factor release the capsaicin receptor from PtdIns(4,5)P2-mediated inhibition. Nature 411: 957–962.

Chuang HH, Yu M, Jan YN, Jan LY (1998) Evidence that the nucleotide exchange and hydrolysis cycle of G-proteins causes acute desensitization of G-protein gated inward rectifier K^+ channels. Proceedings of the National Academy USA 95:11727–11732

Ciani S, Krasne S, Hagiwara S (1980). A model for the effects of potential and external K^+ concentration on the Cs^+ blocking of inward rectification. Biophysical Journal 30: 199–204.

Ciani S, Krasne S, Miyazaki S, Hagiwara S (1978). A model for anomalous rectification: electrochemical-potential-dependent gating of membrane channels. Journal of Membrane Biology 44: 103–134.

Clapham DE (1994) Direct G protein activation of ion channels? Annual Reviews of Neuroscience17:441–464

Clapham DE, Neer EJ (1997) G protein $\beta\gamma$ subunits. Annu Rev Pharmacol Toxicol 37:167–203

Clarson LH, Greenwood SL, Mylona P, Sibley CP (2001). Inwardly rectifying K^+ current and differentiation of human placental cytotrophoblast cells in culture. Placenta 22: 328–336.
Cleemann L, Morad M (1979). Potassium currents in frog ventricular muscle: evidence from voltage clamp currents and extracellular K accumulation. Journal of Physiology 286: 113–143.
Clement IV JP, Kunjilwar K, Gonzalez G, Schwanstecher M, Panten U, Aguilar-Bryan L, Bryan J (1997). Association and stoichiometry of K(ATP) channel subunits. Neuron 18: 827–838.
Codina J, Yatani A, Grenet D, Brown AM, Birnbaumer L (1987). The α subunit of the GTP binding protein G_K opens atrial potassium channels. Science 236:442–445
Cohen NA, Brenman JE, Snyder SH, Bredt DS (1996a). Binding of the inward rectifier K^+ channel Kir2.3 to PSD-95 is regulated by protein kinase A phosphorylation. Neuron 17: 759–767.
Cohen NA, Sha Q, Makhina EN, Lopatin AN, Linder ME, Snyder SH, Nichols CG (1996b). Inhibition of an inward rectifier potassium channel (Kir2.3) by G-protein betagamma subunits. Journal of Biological Chemistry 271: 32301–32305.
Cole KS, Curtis HJ (1941). Membrane potential of the squid giant axon during current flow. Journal of General Physiology 24:551–563
Collins A, Chuang H, Jan YN, Jan LY (1997). Scanning mutagenesis of the putative transmembrane segments of Kir2.1, an inward rectifier potassium channel. Proceedings of the National Academy of Sciences USA 94: 5456–5460.
Collins A, German MS, Jan YN, Jan LY, Zhao B (1996). A strongly inwardly rectifying K^+ channel that is sensitive to ATP Journal of Neuroscience 16: 1–9.
Colquhoun D, Sakmann B (1981) Fluctuations in the microsecond time range of the current through single acetylcholine receptor ion channels. Nature 294:464–466
Colquhoun D, Sakmann B (1985). Fast events in single-channel currents activated by acetylcholine and its analogues at the frog muscle end-plate. Journal of Physiology 369:501–557
Constanti A, Galvan M (1983). Fast inward-rectifying current accounts for anomalous rectification in olfactory cortex neurones. Journal of Physiology 335:153–178
Costantin L L (1970) The role of sodium current in the radial spread of contraction in frog muscle fibers. Journal of General Physiology 55: 703–715.
Corey S, Krapivinsky G, Krapivinsky L, Clapham DE (1998). Number and stoichiometry of subunits in the native atrial G-protein-gated K^+ channel, IK_{ACh}. Journal of Biological Chemistry 273:5271–5278
Coulter KL, Perier F, Radeke CM, Vandenberg CA (1995). Identification and molecular localization of a pH-sensing domain for the inward rectifier potassium channel HIR. Neuron 15: 1157–1168.
Crank J (1967) The mathematics of diffusion. Oxford University Press, London
Crawford AC, Fettiplace R (1980) The frequency selectivity of auditory nerve fibres and hair cells in the cochlea of the turtle. Journal of Physiology 306: 377–412
Dart C, Leyland ML (2001). Targeting of an A-kinase anchoring protein, AKAP79, to an inwardly rectifying potassium channel, Kir2.1. Journal of Biological Chemistry 276: 20499–20505.
Dart C, Leyland ML, Barrett-Jolley R, Shelton PA, Spencer PJ, Conley EC, Sutcliffe MJ, Stanfield PR. (1998a). The dependence of Ag^+ block of a potassium channel, murine Kir2.1, on a cysteine residue in the selectivity filter. Journal of Physiology 511: 15–24.

Dart C, Leyland ML, Spencer PJ, Stanfield PR, Sutcliffe MJ (1998b). The selectivity filter of a potassium channel, murine Kir2.1, investigated using scanning cysteine mutagenesis. Journal of Physiology 511: 25–32.

Dascal N (1997) Signalling via the G protein-activated K^+ channels. Cell Signaling 9: 551–573.

Dascal N, Doupnik CA, Ivanina T, Bausch S, Wang W, Lin C, Garvey J, Chavkin C, Lester HA, Davidson N (1995). Inhibition of function in *Xenopus* oocytes of the inwardly rectifying G-protein-activated atrial K channel (GIRK1) by overexpression of a membrane-attached form of the C-terminal tail. Proceedings of the National Academy USA 92:6758–6762

Dascal N, Schreibmmayer W, Lim NF, Wang W, Chavkin C, DiMagno L, Labarca C, Kieffer BL, Gaveriaux-Ruff C, Trollinger D, Lester HA, Davidson N (1993). Atrial G protein-activated K^+ channel: expression cloning and molecular properties. Proceedings of the National Academy USA 90: 10235–10239.

Davies NW, Shelton PA, Conley EC, Stanfield PR (1996). Subconductance states of the inward rectifier Kir2.1 expressed in murine erythroleukaemia (MEL) cells. Journal of Physiology 495: 89P.

Dell'Acqua ML, Scott JD (1997). Protein kinase A anchoring. Journal of Biological Chemistry 272: 12881–12884.

Demo SD, Yellen G (1991). The inactivation gate of the Shaker K^+ channel behaves like an open channel blocker. Neuron 7: 743–753.

DePaoli AM, Bell GI, Stoffel M (1994) G protein-activated inwardly rectifying potassium channel (GIRK1/KGA) mRNA in adult rat heart and brain by *in situ* hybridization histochemistry. Molecular and Cellular Neurosciences 5:515–522

Derst C, Hirsch JR, Preisig-Müller R, Wischmeyer E, Karschin A, Döring F, Thomzig A, Veh RW, Schlatter E, Kummer W, Daut J (2001a). Cellular localization of the potassium channel Kir7.1 in guinea pig and human kidney. Kidney International 59: 2197–2205.

Derst C, Karschin C, Wischmeyer E, Hirsch JR, Preisig-Muller R, Engel H, Grzeschik K, Daut J, Karschin A (2001b). Genetic and functional linkage of Kir5.1 and Kir2.1 channel subunits. FEBS Letters 491: 205–311.

Dissmann E, Wischmeyer E, Spauschus A, Pfeil DV, Karschin C, Karschin A (1996) Functional expression and cellular mRNA localization of a G protein- activated K^+ inward rectifier isolated from rat brain. Biochemical and Biophysical Research Communications 223: 474–479.

Doi T, Fakler B, Schultz JH, Schulte U, Brandle U, Weidemann S, Zenner HP, Lang F, Ruppersberg JP (1996). Extracellular K^+ and intracellular pH allosterically regulate renal Kir1.1 channels. Journal of Biological Chemistry 271: 17261–17266.

Dong K, Xu J, Vanoye CG, Welch R, MacGregor GG, Giebische G, Hebert SC (2001). An amino acid triplet in the NH_2-terminus of rat ROMK1 determines interaction with SUR2B. Journal of Biological Chemistry 276: 44347–44353.

Döring F, Derst C, Wischmeyer E, Karschin C, Schneggenburger R, Daut J, Karschin A. (1998) The epithelial inward rectifier channel Kir7.1 displays unusual K^+ permeation properties. Journal of Neuroscience 18: 8625–8636.

Doupnik CA, Davidson N, Lester HA (1995a). The inward rectifier potassium channel family. Current Opinion in Neurobiology 5: 268–277.

Doupnik CA, Davidson N, Lester HA, Kofuji P (1997) RGS proteins reconstitute the rapid gating kinetics of Gβγ-activated inwardly rectifying K^+ channels. Proceedings of the National Academy USA 94:10461–10466

Doupnik CA, Dessauer CW, Slepak VZ, Gilman AG, Davidson N, Lester HA (1996) Time resolved kinetics of direct $G_{\beta 1\gamma 2}$ interactions with the carboxyl terminus of Kir3.4 inward rectifier K^+ channel subunits. Neuropharmacology 35:923–931

Doupnik CA, Lim NF, Kofuji P, Davidson N, Lester HA (1995b) Intrinsic gating properties of a cloned G protein-activated inward rectifier K^+ channel. Journal of General Physiology 106:1–23

Doyle DA, Cabral JM, Pfuetzner RA, Kuo A, Gublis JM, Cohen SL, Chait BT, MacKinnon R (1998) The structure of a potassium channel: molecular basis of K^+ conduction and selectivity. Science 280: 69–77.

Drain P, Li L, Wang J (1998). K_{ATP} channel inhibition by ATP requires distinct functional domains of the cytoplasmic C terminus of the pore forming subunit. Proceedings of the National Academy of Sciences USA 95: 13953–13985.

Duprat F, Guillemare E, Romey G, Fink M, Lesage F, Lazdunski M, Honoré E (1995a). Susceptibility of cloned K^+ channels to reactive oxygen species. Proceedings of the National Academy USA 92: 11796–11800.

Duprat F, Lesage F, Guillemare E, Fink M, Hugnot JP, Bigay J, Lazdunski M, Romey G, Barhanin J (1995b) Heterologous mutimeric assembly is essential for K^+ channel activity of neuronal and cardiac G-protein-activated inward rectifiers. Biochemical and Biophysical Research Communications 212:657–663

Edwards FR, Hirst GDS (1988a). Inward rectification in submucosal arterioles of guinea-pig ileum. Journal of Physiology 404: 437–454.

Edwards FR, Hirst GDS, Silverberg GD (1988b). Inward rectification in rat cerebral arterioles; Involvement of potassium ions in autoregulation. Journal of Physiology 404: 455–466.

Edwards G, Dora KA, Gardener MJ, Garland CJ, Weston AH (1998). K^+ is an endothelium-derived hyperpolarizing factor in rat arteries. Nature 396: 269–272.

Einhorn LC, Gregerson KA, Oxford GS (1991) D_2 dopamine receptor activation of potassium channels in identified rat lactotrophs: whole-cell and single-channel recording. Journal of Neurosciences 11: 3727–3737

Einhorn LC, Oxford GS (1993) Guanine nucleotide binding proteins mediate D_2 dopamine receptor activation of a potassium channel in rat lactotrophs. Journal of Physiology 462:563–578

Endoh M, Maruyama M, Iijima T (1985) Attenuation of muscarinic cholinergic inhibition by islet-activating protein in the heart. American Journal of Physiology 249:H309–H320

Escobar L, Root MJ, MacKinnon R (1993). Influence of protein surface charge on the bimolecular kinetics of a potassium channel peptide inhibitor. Biochemistry 32: 6982–6987.

Fakler B, Bond C, Adelman JP, Ruppersberg JP (1996a) Heterololigomeric assembly of inward-rectifier K^+ channels from subunits of different families K(ir)2.1 (IRK1) and K(ir)4.1 (BIR10). Pflügers Archiv 433: 77–83.

Fakler B, Brändle U, Bond C, Glowatzki S, König C, Adelman JP, Zenner H-P, Ruppersberg JP (1994a). A structural determinant of differential sensitivity of cloned inward rectifier K^+ channels to intracellular spermine. FEBS Letters 356: 199–203.

Fakler B, Brändle U, Glowatzki S, Weidemann S Zenner H-P, Ruppersberg JP (1995). Strong voltage-dependent inward rectification of inward rectifier K^+ channels is caused by intracellular spermine. Cell 80: 149–154.

Fakler B, Brändle U, Glowatzki E, Zenner HP, Ruppersberg JP (1994b) $K_{ir}2.1$ inward rectifier K^+ channels are regulated independently by protein kinases and ATP hydrolysis. Neuron 13:1413–1420

Fakler B, Schultz JH, Yang J, Schulte U, Brandle U, Zenner HP, Jan LY, Ruppersberg JP (1996b) Identification of a titratable lysine residue that determines sensitivity of kidney potassium channels (ROMK) to intracellular pH. EMBO Journal 15: 4093–4099.

Fan Z, Makielski JC (1997). Anionic phospholipids activate ATP-sensitive potassium channels. Journal of Biological Chemistry 272: 5388–5395.

Farkas RH, Chien PY, Nakajima S, Nakajima Y (1996) Properties of a slow nonselective cation conductance modulated by neurotensin and other neurotransmitters in midbrain dopaminergic neurons Journal of Neurophysiology 76: 1968–1981

Farkas RH, Chien PY, Nakajima S, Nakajima Y (1997) Neurotensin and dopamine D2 activation oppositely regulate the same K^+ conductance in rat midbrain dopaminergic neurons. Neurosciences Letters 231:21–24

Farkas RH, Nakajima S, Nakajima Y (1994) Neurotensin excites basal forebrain cholinergic neurons: ionic and signal-transduction mechanisms. Proceedings of the National Academy USA 91:2853–2857

Ferrer J, Nichols CG, Makhina EN, Salkoff L, Bernstein J, Gerhard D, Wasson J, Ramanadham S, Permutt A (1995) Pancreatic islet cells express a family of inwardly rectifying K^+ channel subunits which interact to form G-protein-activated channels. Journal of Biological Chemistry 270:26086–26091

Ficker E, Taglialatela M, Wible BA, Henley CM, Brown AM (1994). Spermine and spermidine as gating molecules for inward rectifier K^+ channels. Science 266: 1068–1072.

Fink M, Duprat F, Heurteaux C, Lesage F, Romey G, Barhanin J, Lazdunski M (1996). Dominant negative chimeras provide evidence for homo and heteromultimeric assembly of inward rectifier K^+ channel proteins via their N-terminal end. FEBS Letters 378: 64–68.

Fischer-Lougheed J, Liu J-H, Espinos E, Mordasini D, Bader CR, Belin D, Bernheim L (2001). Human myoblast fusion requires expression of functional inward rectifier Kir2.1 channels. Journal of Cell Biology 153: 677–685.

Fletcher JE, Lindorfer MA, DeFilippo JM, Yasuda H, Guilmard M, Garrison JC (1998) The G protein β_5 subunit interacts selectively with the $G_q\alpha$ subunit. Journal of Biological Chemistry 273:636–644

Ford CE, Skiba NP, Bae H, Daaka Y, Reuveny E, Shekter LR, Rosal R, Weng G, Yang CS, Iyengar R, Miller RJ, Jan LY, Lefkowitz RJ, Hamm HE (1998) Molecular basis for interactions of G Protein $\beta\gamma$ subunits with effectors. Science 280:1271–1274

Forsyth SE, Hoger A, Hoger JH (1997). Molecular cloning and expression of a bovine endothelial inward rectifier potassium channel FEBS Letters 409: 277–282.

Freissmuth M, Casey PF, Gilman AG (1989) G proteins control diverse pathways of transmembrane signaling. FASEB Journal 3:2125–2132

French RJ, Shoukimas JJ (1981). Blockage of squid axon potassium conductance by internal tetra-*n*-alkylammonium ions of various sizes. Biophysical Journal 34: 271–291.

Fukushima Y (1982) Blocking kinetics of the anomalous potassium rectifier of tunicate egg studied by single channel recording. Journal of Physiology 331:311–331

Gay LA, Stanfield PR (1977) Cs^+ causes a voltage-dependent block of inward K currents in resting skeletal muscle fibres. Nature 267: 169–170.

Gerber U, Stevens DR, McCarley RW, Greene RW (1991) Muscarinic agonists activate an inwardly rectifying potassium conductance in medial pontine reticular formation neurons of the rat in vitro. Journal of Neurosciences 11:3861–3867

Giebisch G (1998). Renal potassium transport: mechanism and regulation. American Journal of Physiology 274: F817–833.

Glowatzki E, Fakler G, Brandle U, Rexhausen U, Zenner H-P, Ruppersberg JP, Fakler B (1995). Subunit-dependent assembly of inward-rectifier K^+ channels. Proceedings of the Royal Society, B. 261: 251–261.

Goodman MB, Art JJ (1996) Positive feedback by a potassium-selective inward rectifier enhances tuning in vertebrate hair cells. Biophysical Journal 71:430–442

Gosset P, Ghezala GA, Korn B, Yapso ML, Poutska A, Lehrach H, Sinet PM, Creau N (1997) A new inward rectifier potassium channel gene (KCNJ15) localized on chromosome 21 in the Down syndrome chromosome region 1 (DCR1). Genomics 44: 237–241

Grafe P, Rimpel J, Reddy MM, TenBruggencate G (1982) Changes of intracellular sodium and potassium ion concentrations in frog spinal motoneurons induced by repetitive synaptic stimulation. Neuroscience 7: 3213–3220

Grigg JJ, Kozasa T, Nakajima Y, Nakajima S (1996) Single-channel properties of a G-protein-coupled inward rectifier potassium channel in brain neurons. Journal of Neurophysiology 75:318–328

Gulbis JM, Zhou M, Mann S, MacKinnon R (2000). Structure of the cytoplasmic β subunit-T1 assembly of voltage-dependent K^+ channels. Science 289:123–127.

Guo D, Lu Z (2000a). Mechanism of cGMP-gated channel block by intracellular polyamines. Journal of General Physiology 115: 783–797.

Guo D, Lu Z (2000b). Mechanism of IRK1 channel block by intracellular polyamines. Journal of General Physiology 115: 799–813.

Guo D, Lu Z (2000c). Pore block versis intrinsic gating in the mechanism of inward rectification in strongly rectifying IRK1 channels. Journal of General Physiology 116: 561–568.

Guo D, Lu Z (2001). Kinetics of inward-rectifier K^+ channel block by quaternary alkylammonium ions: dimensions and properties of the inner pore. Journal of General Physiology 117: 395–405.

Guo L, Kubo Y (1998). Comparison of the open-close kinetics of the cloned inward rectifier K^+ channel IRK1 and its point mutant (Q140E) in the pore region. Receptors & Channels 5: 273–289.

Hagiwara S, Jaffe LA (1979) Electrical properties of egg cell membranes. Annual Reviews of Biophysics and Bioengineering 8:385–416

Hagiwara S, Miyazaki S, Krasne S, Ciani S (1977). Anomalous permeabilities of the egg cell membrane of a starfish in K^+-Tl^+ mixtures. Journal of General Physiology 70: 269–281.

Hagiwara S, Miyazaki S, Moody W, Patlak J (1978). Blocking effects of Ba and H ions on the K current during anomalous rectification in the starfish egg. Journal of Physiology 279: 167–185.

Hagiwara S, Miyazaki S, Rosenthal NP (1976). Potassium current and the effect of cesium on this current during anomalous rectification of the egg cell membrane of a starfish. Journal of General Physiology 67: 621–638.

Hagiwara S., Takahashi K (1974). The anomalous rectification and cation selectivity of the membrane of a starfish egg. Journal of Membrane Biology 18: 61–80.

Hagiwara S, Yoshii M (1979). Effects of internal potassium and sodium on the anomalous rectification of the starfish egg as examined by internal perfusion. Journal of Physiology 292: 251–265.

Hansen CA, Schroering AG, Carey DJ, Robishaw JD (1994) Localization of a heterotrimeric G protein γ subunit to focal adhesions and associated stress fibers. Journal of Cell Biology 126:811–819

Hardie RC, Raghu P, Moore S, Juusola M, Baines RA, Sweeney ST (2001). Calcium influx via trp channels is required to maintain PIP_2 levels in *Drosophila* photoreceptors. Neuron 30: 149–159.

He C, Zhang H, Mirshahi T, Logothetis DE (1999). Identification of a potassium channel site that interacts with G protein βγ subunits to mediate agonist-induced signaling. Journal of Biological Chemistry 274: 12517–12524.

Hedin KE, Lim NF, Clapham DE (1996) Cloning of a *Xenopus* laevis inwardly rectifying K^+ channel subunit that permits GIRK1 expression of I_{KACh} currents in oocytes. Neuron 16:423–429

CHO HC Tsushima RG, Nguyen T-TT, Guy HR, Backx PH (2000). Two critical cysteine residues implicated in disulfide bond formation and proper folding of Kir2.1. Biochemistry: 39: 4649–4657.

Heginbotham L, Lu Z, Abramson T, MacKinnon R (1994). Mutations in the K^+ channel signature sequence. Biophysical Journal 66:1061–1067.

Heiny JA, Ashcroft FM, Vergara J (1983). T-system optical signals associated with inward rectification in skeletal muscle. Nature 301: 164–166.

Henry P, Pearson WL, Nichols CG (1996). Protein kinase C inhibition of cloned inward rectifier (HRK1/Kir2.3) K^+ channels expressed in *Xenopus* oocytes. Journal of Physiology 495: 681–688.

Herlitze S, Ruppersberg JP, Mark MD (1999) New roles for RGS2, 5, and 8 on the ratio-dependent modulation of recombinant GIRK channels expressed in *Xenopus* oocytes. Journal of Physiology (1999) 517: 341–352.

Hestrin S (1981). The interaction of potassium with the activation of anomalous rectification in frog muscle membrane. Journal of Physiology 317: 497–508.

Hibino H, Inanobe A, Tanemoto M, Fujita A, Doi K, Kubo T, Hata Y, Takai Y, Kurachi Y (2000). Anchoring proteins confer G protein sensitivity to an inward rectifier K^+ channel through the GK domain. EMBO Journal 19: 78–83.

Hilgemann DW, Ball R (1996). Regulation of cardiac Na^+,Ca^{2+} exchanger and K(ATP) potassium channels by PIP_2. Science 273: 956–959.

Hill JJ, Peralta EG (2001) Inhibition of a Gi-activated potassium channel (GIRK1/4) by the Gq-coupled m1 muscarinic acetylcholine receptor. Journal of Biological Chemistry 276: 5505–5510.

Hille B (1973) Potassium channels in myelinated nerve: selective permeability to small cations. Journal of General Physiology 61: 669–686.

Hille B (2001). Ionic Channels of Excitable Membranes 3rd edition. Sinauer, Sunderland Mass. 814pp.

Hille B, Schwarz W (1978). Potassium channels as multi-ion single file pores. Journal of General Physiology 72: 409–442.

Ho IHM, Murrell-Lagnado RD (1999a). Molecular mechanism for sodium-dependent activation of G protein-gated K^+ channels. Journal of Physiology 520: 645–651.

Ho IHM, Murrell-Lagnado RD (1999b) Molecular determinants for sodium-dependent activation of G protein-gated K^+ channels. Journal of Biological Chemistry 274: 8639–8648.

Ho K, Nichols CG, Lederer WJ, Lytton J, Vassilev PM, Kanazirska MV, Hebert SC (1993) Cloning and expression of an inwardly rectifying ATP-regulated potassium channel. Nature 362: 31–38.

Hodgkin AL (1992) Chance & Design: reminiscences of science in peace and war. Cambridge University Press 412pp.

Hodgkin A L, Horowicz P (1959a) Movements of Na and K in single muscle fibres. Journal of Physiology 145: 405–432

Hodgkin AL, Horowicz P (1959b). The influence of potassium and chloride ions on the membrane potential of single muscle fibres. Journal of Physiology 148: 127–160.

Hodgkin AL, Horowicz P (1960). The effect of sudden changes in ionic concentration on the membrane potential of single muscle fibres. Journal of Physiology 153: 370–385.

Hodgkin AL, Huxley AF, Katz B (1949) Ionic currents underlying activity in the giant axon of the squid. Archives des Sciences Physiologiques 3:129–150

Hodgkin AL, Keynes RD (1955). The potassium permeability of a giant nerve fibre. Journal of Physiology 128: 61–88.

Horie M, Irisawa H (1987) Rectification of muscarinic K^+ current by magnesium ion in guinea pig atrial cells. American Journal of Physiology 253:H210–214

Horie M, Irisawa H (1989) Dual effects of intracellular magnesium on muscarinic potassium channel current in single guinea-pig atrial cells. Journal of Physiology 408:313–332

Horio Y, Hibino H, Inanobe A, Yamada M, Ishii M, Tada Y, Satoh E, Hata Y, Takai Y, Kurachi Y (1997). Clustering and enhanced activity of an inwardly rectifying potassium channel, Kir4.1, by an anchoring protein, PSD-95/SAP90. Journal of Biological Chemistry 272: 12885–12888

Horio Y, Kurachi Y (1999). Glial inwardly rectifying potassium channels. Current Topics in Membranes 46: 471– 484.

Hoshi T, Zagotta WN, Aldrich RW (1990). Biophysical and molecular mechanisms of Shaker potassium channel inactivation. Science 250: 533–538.

Hou P, Yan S, Tang W, Nelson DJ (1999) The inwardly rectifying K^+ channel subunit GIRK1 rescues the GIRK2 *weaver* phenotype. Journal of Neuroscience 19: 8327–8336.

Huang C-J, Moczydlowski E (2001). Cytoplasmic polyamines as permeant blockers and modulators of the voltage-gated sodium channel. Biophysical Journal 80: 1262–1279.

Huang CL, Fen S, Hilgemann DW (1998). Direct activation of inward rectifier potassium channels by PIP_2 and its stabilisation by Gbetagamma, Nature 391: 803–806.

Huang CL, Jan YN, Jan LY (1997) Binding of the G protein $\beta\gamma$ subunit to multiple regions of G protein-gated inward-rectifying K^+ channels. FEBS Lett 405:291–298

Huang CL, Slesinger PA, Casey PJ, Jan YN, Jan LY (1995) Evidence that direct binding of $G_{\beta\gamma}$ to the GIRK1 G protein-gated inwardly rectifying K^+ channel is important for channel activation. Neuron 15:1133–1143

Hughes BA, Kumar G, Yuan Y, Swaminathan A, Yan D, Sharma A, Plumley L, Yang-Feng TL, Swaroop A (2000). Cloning and functional expression of human retinal Kir2.4, a pH-sensitive inwardly rectifying K^+ channel. American Journal of Physiology 279: C771–C784.

Ibarra J, Morley GE, Delmar M (1991). Dynamics of the inward rectifier K^+ current during the action potential of guinea pig ventricular myocytes. Biophysical Journal 60: 1534–1539.

Inagaki N, Gonoi T, Clement IV JP, Namba N, Inazawa J, Gonzalez G, Aguilar-Bryan L, Seino S, Bryan J (1995a) Reconstitution of $I(K_{ATP})$; an inward rectifier subunit plus the sulfonylurea receptor. Science 270: 1166–1170.

Inagaki N, Gonoi T, Seino S (1997). Subunit stoichiometry of the pancreatic beta-cell ATP-sensitive K^+ channel. FEBS Letters 409: 232–236.

Inagaki N, Tsuura Y, Namba N, Masuda K, Gonoi T Horie M, Seino Y, Mizuta M, Seino S (1995b) Cloning and functional characterization of a novel ATP-sensitive potassium channel ubiquitously expressed in rat tissues, including pancreatic islets, skeletal muscle, and heart. Journal of Biological Chemistry 270: 5691–5694.

Inanobe A, Horio Y, Fujita A, Tanemoto M, Hibino H, Inageda K, Kurachi Y (1999) Molecular cloning and characterization of a novel splicing variant of the Kir3.2 subunit predominantly expressed in mouse testis. Journal of Physiology 521:19–30.

Inanobe A, Ito H, Ito M, Hosoya Y, Kurachi Y (1995a) Immunological and physical characterization of the brain G protein-gated muscarinic potassium channel. Biochemical and Biophysical Research Communications 217:1238–1244

Inanobe A, Morishige KI, Takahashi N, Ito H, Yamada M, Takumi T, Nishina H, Takahashi K, Kanaho Y, Katada T, Kurachi Y (1995b) $G_{\beta\gamma}$ directly binds to the carboxyl terminus of the G protein-gated muscarinic K^+ channel, GIRK1. Biochemical and Biophysical Research Communications 212:1022–1028

Iñiguez-Lluhi JA, Simon MI, Robishaw JD, Gilman AG (1992) G protein βγ subunits synthesized in Sf9 cells. Journal of Biological Chemistry 267:23409–23417

Inoue M, Nakajima S, Nakajima Y (1988) Somatostatin induces an inward rectification in rat locus coeruleus neurones through a pertussis toxin-sensitive mechanism. Journal of Physiology 407:177–198

Isa T, Lino M, Itazawa S-I, Ozawa S (1995). Spermine mediates inward rectification of Ca^{2+}-permeable AMPA receptor channels. Neuroreport 6: 2045–2048.

Ishihara K (1997). Time dependent outward currents through the inward rectifier potassium channel IRK1. Journal of General Physiology 109: 229–243.

Ishihara K, Ehara T (1998). A repolarization-induced transient increase in the outward current of the inward rectifier K^+ channel in guinea-pig cardiac myocytes. Journal of Physiology 510: 755–771.

Ishihara K, Hiraoka M, Ochi R (1996). The tetravalent organic cation spermine causes the gating of the IRK1 channel expressed in murine fibroblast cells. Journal of Physiology 491: 367–381.

Ishihara K, Mitsuiye T, Noma A, Takano M (1989). The Mg^{2+} block and intrinsic gating underlying inward rectification of the K^+ current in guinea-pig cardiac myocytes. Journal of Physiology 419: 297–320.

Ishii K, Yamagishi T, Taira N (1994). Cloning and functional expression of a cardiac inward rectifier K^+ channel. FEBS Letters 338: 107–111.

Ishii M, Horio Y, Tada Y, Hibino H, Inanobe A, Ito M, Yamada M, Gotow T, Uchiyama Y, Kurachi Y (1997) Expression and clustered distribution of an inwardly rectifying potassium channel, K(AB)-2/Kir4.1, on mammalian retinal Muller cell membrane: their regulation by insulin and laminin signals. Journal of Neuroscience 17: 7725–7735.

Isomoto S, Kondo C, Kurachi Y (1997) Inwardly rectifying potassium channels: Their molecular heterogeneity and function.Japanese Journal of Physiology 47:11–39

Isomoto S, Kondo C, Takahashi N, Matsumoto S, Yamada M, Takumi T, Horio Y, Kurachi Y (1996) A novel ubiquitously distributed isoform of GIRK2 (GIRK2B) enhances GIRK1 expression of the G-protein-gated K+ current in *Xenopus* oocytes. Biochemical and Biophysical Research Communications 218:286–291

Ito H, Tung RT, Sugimoto T, Kobayashi I, Takahashi K, Katada T, Ui M, Kurachi Y (1992) On the mechanism of G protein βγ subunit activation of the muscarinic K^+ channel in guinea pig atrial cell membrane. Journal of General Physiology 99:961–983

Ivanova-Nikolova TT, Breitwieser GE (1997) Effector contributions to Gβγ-mediated signaling as revealed by muscarinic potassium channel gating. Journal of General Physiology 109:245–253

Ivanova-Nikolova TT, Nilolov EN, Hansen C, Robishaw JD (1998) Muscarinic K^+ channel in the heart. Modal regulation by G protein beta gamma subunits. J Gen Physiol 112: 199–210.

Jackson MB, Wong BS, Morris CE, Lecar H (1983) Successive openings of the same acetylcholine receptor channel are correlated in open time. Biophysical Journal 42:109–114

Jan LY, Jan YN (1982) Peptidergic transmission in sympathetic ganglia of the frog. Journal of Physiology 327:219–246

Jan LY, Jan YN (1994) Potassium channels and their evolving gates. Nature 371:119–122

Jan LY, Jan YN (1997) Cloned potassium channels from eukaryotes and prokaryotes. Annual Reviews of Neuroscience 20:91–123

Janmey PA, Xian W, Flanagan LA (1999). Controlling cytoskeletal structure by phosphoinositide-protein interactions: phosphoinositide binding protein domains and effects of lipid packing. Chemistry and Physics of Lipids 101: 93–107.

Jelacic TM, Kennedy ME, Wickman K, Clapham DE (2000) Functional and biochemical evidence for G-protein-gated inwardly rectifying K^+ (GIRK) channels composed of GIRK2 and GIRK3. Journal of Biological Chemistery 275: 36211–36216.

Jelacic TM, Sims SM, Clapham, DE (1999) Functional expression and characterization of G-protein-gated inwardly rectifying K^+ channels containing GIRK3. Journal Membrane Biologn 169: 123–129.

Jiang ZG, Pessia M, North RA (1994) Neurotensin excitation of rat ventral tegmental neurones. Journal of Physiology 474:119–129

John SA, Monck JR, Weiss JN, Ribalet B (1998). The sulphonylurea receptor SUR1 regulates ATP-sensitive mouse Kir6.2 K^+ channel linked to the green fluorescent protein in human embryonic kidney cells (HEK 293). Journal of Physiology 510: 333–345.

Jones SVP (1996). Modulation of the inwardly rectifying potassium channel IRK1 by the m1 muscarinic receptor. Molecular Pharmacology 49: 662–667.

Jones SVP (1997). Dual modulation of an inwardly rectifying potassium conductance. Neuropharmacology 36: 209–215.

Kandel ER, Tauc L (1966). Anomalous rectification in the metacerebral giant cells and its concequences for synaptic transmission. Journal of Physiology 183: 287–304.

Kang JX, Xiao Y-F, Leaf A (1995). Free, long-chain, polyunsaturated fatty acids reduce membrane electrical excitability in neonatal rat cardiac myocytes. Proceedings of the National Academy of Sciences USA 92: 3997–4001.

Karschin A, Wischmeyer E (1995) Identification of G protein-regulated inwardly rectifying K^+ channels in rat brain oligodendrocytes. Neurosci Lett 183:135–138

Karschin A, Wischmeyer E, Davidson N, Lester HA (1994). Fast inhibition of inwardly rectifying K^+ channels by multiple neurotransmitter receptors in oligodendroglia. European Journal of Neuroscience 6: 1756–1764.

Karschin C, DiBmann E, Stühmer W, Karschin A (1996) IRK(1–3) and GIRK(1–4) inwardly rectifying K^+ channel mRNAs are differentially expressed in the adult rat brain. Journal of Neuroscience 16:3559–3570

Karschin C, Karschin A (1997). Ontogeny of gene expression of Kir channel subunits in the rat. Molecular & Cellular Neurosciences 10: 131–148.

Karschin C, Schreibmayer W, Dascal N, Lester H, Davidson N, Karschin A (1994) Distribution and localization of a G protein-coupled inwardly rectifying K^+ channel in the rat. FEBS Letters 348:139–144.

Katz B (1949). Les constantes électriques de la membrane du muscle. Archives des Sciences Physiologiques 3: 285–299.

Kennedy ME, Nemec J, Corey S, Wickman K, Clapham DE (1999). GIRK4 confers appropriate processing and cell surface localization to G-protein gated potassium channels. Journal of Biological Chemstry 274: 2571–2582.

Kerkut GA, Thomas RC (1965) An electrogenic sodium pump in snail nerve cells. Comparative Biochemistry and Physiology 14: 167–183

Kim D (1991) Modulation of acetylcholine-activated K^+ channel function in rat atrial cells by phosphorylation. Journal of Physiology 437:133–155.

Kim D, Bang H (1999) Modulation of rat atrial G protein-coupled K^+ channel function by phospholipids. Journal of Physiology 517: 59–74.

Kim E, Niethammer M, Rothschild A, Jan YN, Sheng M (1995). Clustering of Shaker-type K^+ channels by interaction with a family of membrane-associated guanylate kinases. Nature 378: 85–88.

Kim KM, Nakajima S, Nakajima Y (1997) Dopamine and GABA receptors in cultured substantia nigra neurons: correlation of electrophysiology and immunocytochemistry. Neuroscience 78: 759–769

Kim KM, Nakajima Y, Nakajima S (1995) G protein-coupled inward rectifier activated by dopamine agonists in cultured substantia nigra neurons. Neuroscience 69:1145–1158

Kim SJ, Kerst G, Schreiber R, Pavenstadt H, Greger R, Hug MJ, Bleich M (2000). Inwardly rectifying K+ channels in the basolateral membrane of rat pancreatic acini. Pflügers Archiv 441:331–340.

Kirsch GE, Nichols RA, Nakajima S. (1977) Delayed rectification in the transverse tubules. Origin of the late after-potential in frog skeletal muscle. Journal of General Physiology 70: 1–21.

Kisselev O, Gautam N (1993) Specific interaction with rhodopsin is dependent on the g subunit type in a G protein.Journal of Biological Chemistry 268:24519–24522

Kleuss C, Hescheler J, Ewel C, Rosenthal W, Schultz G, Wittig B (1991) Assignment of G-protein subtypes to specific receptors inducing inhibition of calcium currents. Nature 353:43–48

Kleuss C, Scherubl H, Hescheler J, Schultz G, Wittig B (1992) Different β-subunits determine G-protein interaction with transmembrane receptors. Nature 358:424–426

Kleuss C, Scherubl H, Hescheler J, Schultz G, Wittig B (1993) Selectivity in signal transduction determined by β subunits of heterotrimeric G proteins. Science 259:832–834

Knot HJ, Zimmermann PA, Nelson MT (1996). Extracellular K^+-induced hyperpolarizations and dilatations of rat coronary and cerebral arteries involve inward rectifier K^+ channels. Journal of Physiology 492: 419–430.

Kobayashi T, Ikeda K, Ichikawa T, Abe S, Togashi S, Kumanishi T (1995) Molecular cloning of a mouse G-protein-activated K^+ channel (mGIRK1) and distinct distributions of three GIRK (GIRK1, 2 and 3) mRNAs in mouse brain. Biochemical and Biophysical Research Communications 208:1166–1173

Kobertz WR, Williams C, Miller C (2000) Hanging gondola structure of the T1 domain in a voltage gated K^+ channel. Biochemistry 39: 10437–10352.

Kobrinsky E, Mirshahi T, Zhang H, Jin T, Logothetis DE (2000) Receptor-mediated hydrolysis of plasma membrane messenger PIP_2 leads to K^+-current desensitization. Nature Cell Biologn 2: 507–514.

Kofuji P, Davidson N, Lester HA (1995) Evidence that neuronal G-protein-gated inwardly rectifying K^+ channels are activated by Gβγ subunits and function as heteromultimers. Proceedings of the National Academy of Sciences USA 92:6542–6546

Kofuji P, Doupnik CA, Davidson N, Lester HA (1996a) A unique P-region residue is required for slow voltage-dependent gating of a G protein-activated inward rectifier K^+ channel expressed in *Xenopus* oocytes. Journal of Physiology 490.3:633–645

Kofuji P, Hofer M, Millen KJ, Millonig JH, Davidson N, Lester HA, Hatten M (1996b). Functional analysis of the weaver mutant GIRK2 K^+ channel and rescue of *weaver* granule cells. Neuron 16: 941–952.

Kondo C, Isomoto S, Matsumoto S, Yamada M, Horio Y, Yamashita S, Takemura-Kameda K, Matsuzawa Y, Kurachi Y (1996) Cloning and functional expression of a novel isoform of ROMK inwardly rectifying ATP-dependent K^+ channel, ROMK6 (Kir1.1f). FEBS Letters 399:122–126

Korchev YE, Bashford CL, Alder GM, Apel PY, Edmonds DT, Lev AA, Nandi K, Zima AV, Pasternak CA (1997) A novel explanation for fluctuations of ion current through narrow pores. FASEB Journal 11: 600–608.
Koumi SI, Wasserstrom JA, Ten Eick RE (1995a) β-adrenergic and cholinergic modulation of the inwardly rectifying K^+ current in guinea-pig ventricular myocytes. Journal of Physiology 486:647–659
Koumi SI, Wasserstrom JA, Ten Eick RE (1995b) β-adrenergic and cholinergic modulation of inward rectifier K^+ channel function and phosphorylation in guinea-pig ventricle. Journal of Physiology 486:661–678
Kovoor, A., Chen, C-K, He, W., Wensel, T.G., Simon, M.I., and Lester H.A . (2000) Co-expression of Gβ5 enhances the function of two Gγ subunit-like domain-containing regulators of G protein signaling proteins. Journal of Biological Chemistry **275:** 3397–3402.
Koyama H, Morishige K-I, Takahashi N, Zanelli JS, Fass DN, Kurachi Y (1994). Molecular cloning, functional expression and localization of a novel inward rectifier potassium channel in the rat brain. FEBS Letters 341: 303–307.
Koyano K, Velimirovic BM, Grigg JJ, Nakajima S, Nakajima Y (1993) Two signal transduction mechanisms of substance P-induced depolarization in locus coeruleus neurons. European Journal of Neuroscience 5: 1189–1197
Krapivinsky G, Gordon EA, Wickman B, Velimirovic B, Krapivinsky L, Clapham DE (1995a). The G protein gated atrial K^+ channel IK(Ach) is a hetermomultimer of 2 inwardly rectifying K^+ channel proteins. Nature 374: 135–141.
Krapivinsky G, Kennedy ME, Nemec J, Medina I, Krapivinsky L, Clapham DE (1998a) Gβγ binding to GIRK4 subunit is critical for G protein-gated K^+ channel activation. Journal of Biological Chemistry 273: 16946–16952
Krapivinsky G, Krapivinsky L, Wickman K, Clapham DE (1995b) Gβγ binds directly to the G protein-gated K^+ channel, I_{KACh}. Journal of Biological Chemistry 270:29059–29062
Krapivinsky G, Medina I, Eng L, Krapivinsky L, Yang Y, Clapham DE (1998b). Novel inward rectifier K^+ channel with unique pore properties. Neuron 20: 995–1005.
Kuba K, Koketsu K (1976) Analysis of the slow excitatory postsynaptic potential in bullfrog sympathetic ganglion cells. Japanese Journal of Physiology 26:651–669
Kubo Y (1996). Effects of extracellular cations and mutations in the pore region of the inward rectifier K^+ channel IRK1. Receptors & Channels 4: 73–83.
Kubo Y, Baldwin TJ, Jan YN & Jan LY (1993a) Primary structure and functional expression of a mouse inward rectifier potassium channel. Nature 362:127–133.
Kubo Y, Miyashita T, Kubokawa K (1996) A weakly inward rectifying potassium channel of the salmon brain. Glutamate 179 in the second transmembrane domain is insufficient for strong rectification. Journal of Biological Chemistry 271: 15729–15735.
Kubo Y, Murata Y (2001). Control of rectification and permeation by two distinct sites after the second transmembrane region in Kir2.1 K^+ channel. Journal of Physiology 531: 645–660.
Kubo Y, Reuveny E, Slesinger PA, Jan YN & Jan LY (1993b) Primary structure and functional expression of a rat G-protein coupled muscarinic potassium channel. Nature 364:802–806.
Kubo Y, Yoshimichi M, Heinemann SH (1998). Probing pore topology and conformational changes of Kir2.1 potassium channels by cysteine scanning mutagenesis. FEBS Letters 435: 69–73.
Kuffler SW, Sejnowski TJ (1983) Peptidergic and muscarinic excitation at amphibian sympathetic synapses. Journal of Physiology 341:257–278

Kumpf RA Dougherty DA (1993). A mechanism for ion selectivity in potassium channels: computational studies of cation-p interactions. Science 261: 1708–1710.

Kunkel MT, Peralta EG (1995) Identification of domains conferring G protein regulation on inward rectifier potassium channels. Cell 83:443–449

Kurachi Y (1985). Voltage-dependent activation of the inward rectifier potassium channel in the ventricular cell membrane of guinea-pig heart. Journal of Physiology 366: 365–385.

Kurachi Y, Nakajima T, Sugimoto T (1986) On the mechanism of activation of muscarinic K^+ channels by adenosine in isolated atrial cells: involvement of GTP-binding proteins. Pflügers Archiv 407:264–274

Kurachi Y, Nakajima T, Sugimoto T (1987) Short-term desensitization of muscarinic K^+ channel current in isolated atrial myocytes and possible role of GTP-binding proteins. Pflügers Archiv 410: 227–233

Kurschner C, Mermelstein PG, Holden WT, Surmeier DJ (1998). CIPP, a novel multivalent PDZ domain protein, selectively interacts with Kir4.0 family members, NMDA receptor subunits, neurexins, and neuroligins. Molecular & Cellular Neurosciences 11: 161–172.

Kürz LL, Kuhlke RD, Zhang HJ, Joho RH (1995). Side chain accessibilities in the pore of a K channel probed by sulfhyrdyl reagents after scanning cysteine mutagenesis. Biophysical Journal 68: 900–905.

Kusaka S, Inanobe A, Fujita A, Makino Y, Tanemoto M, Matsushita K, Tano Y, Kurachi Y (2001). Functional Kir7.1 channels localized at the root of apical processes in rat retinal pigment epithelium. Journal of Physiology 531: 27–36.

Kuschinsky W, Wahl M, Bosse O, Thurau K (1972) Perivascular potassium and pH as determinants of local pial arterial diameter in cats. Circulation Research 31: 240–247.

Kwon G, Axelrod D, Neubig RR (1994) Lateral mobility of tetramethylrhodamine (TMR) labelled G protein α and βγ subunits in NG 108–15 cells. Cellular Signalling 6:663–679

Kyte J, Doolittle RF (1982). A simple method for displaying the hydropathic character of a protein. Journal of Molecular Biology 157: 105–132.

Lagrutta AA, Bond CT, Xia XM, Pessia M, Tucker S, Adelman JP (1996) Inward rectifier potassium channels:cloning, expression and structure-function studies. Japanese Heart Journal 37: 651–660.

Lambright DB, Sondek J, Bohm A, Skiba NP, Hamm HE, Sigler PB (1996) The 2.0 Å crystal structure of a heterotrimeric G protein. Nature 379:311–319

Lancaster MK, Dibb KM, Quinn CC, Leach R, Lee J-K, Findlay JBC, Boyett MR (2000). Residues and mechanisms for slow activation and Ba^{2+} block of the cardiac muscarinic K^+ channel, Kir3.1/Kir3.4. Journal of Biological Chemistry 275: 35831–35839.

Leaney JL, Dekker LV, Tinker A (2001) Regulation of a G protein-gated inwardly rectifying K^+ channel by a Ca^{2+}-independent protein kinase C. Journal of Physiology 534: 367–379.

Lee J-K, John SA, Weiss JN (1999). Novel gating mechanism of polyamine block in the strong inward rectifier K channel Kir2.1. Journal of General Physiology 113: 555–563.

Leech CA, Stanfield PR (1981) Inward rectification in frog skeletal muscle fibres and its dependence on membrane potential and external potassium. Journal of Physiology 319: 295–309.

Lei Q, Jones MB, Talley EM, Schrier AD, McIntire, WE, Garrison JC, and Bayliss DA (2000) Activation and inhibition of G protein-coupled inwardly rectifying potassium (Kir3) channels by G protein βγ subunits. Proceedings of the National Academy of Sciences USA 97: 9771–9776.

Lei Q, Talley EM, Bayliss DA (2001) Receptor-mediated inhibition of G protein-coupled inwardly rectifying potassium channels involves $G\alpha_q$ family subunits, phospholipase C, and a readily diffusible messenger. Journal of Biological Chemistry 276: 16720–16730.

Leonoudakis D, Mailliard WS, Wingerd KL, Clegg DO, Vandenberg CA (2001). Inward rectifier potassium channel Kir2.2 is associated with synapse-associated protein SAP97. Journal of Cell Science 114: 987–998.

Lesage F, Duprat F, Fink M, Guillemare E, Coppola T, Lazdunski M, Hugnot JP (1994) Cloning provides evidence for a family of inward rectifier and G-protein coupled K^+ channels in the brain. FEBS Letters 353:37–42

Lesage F, Guillemare E, Fink M, Duprat F, Heurteaux C, Fosset M, Romey G, Barhanin J, Lazdunski M (1995) Molecular properties of neuronal G-protein-activated inwardly rectifying K^+ channels. Journal of Biological Chemistry 270:28660–28667

Leyland ML, Dart C, Spencer PJ, Sutcliffe MJ, Stanfield PR (1999). The possible role of a disulphide bond in forming functional Kir2.1 channels. Pflügers Archiv 438: 778–781.

Li M, Jan YN, Jan LY (1992). Specification of subunit assembly by the hydrophilic amino terminal domain of the Shaker potassium channel. Science 257: 1225–1230.

Li M, Unwin N, Stauffer KA, Jan YN, Jan LY (1994). Images of purified *Shaker* potassium channels. Current Biology 4: 110–115.

Li Y, Sternweis PM, Charnecki S, Smith TF, Gilman AG, Neer EJ, Kozasa T (1998) Sites for Gα binding on the G protein β subunit overlap with sites for regulation of phospholipase Cβ and adenylyl cyclase. Journal of Biological Chemistry 273:16265–16272

Liao YJ, Jan YN, Jan LY (1996) Heteromultimerization of G protein-gated inwardly rectifying K^+ channel proteins GIRK1 and GIRK2 and their altered expression in *weaver* brain. Journal of Neuroscience 16:7137–7150.

Lim NF, Dascal N, Labarca C, Davidson N, Lester HA (1995) A G protein-gated K channel is activated via β2-adrenergic receptors and Gβγ subunits in *Xenopus* oocytes. Journal of General Physiology 105:421–439

Liou HH, Zhou SS, Huang CL (1999). Regulation of ROMK1 channel by protein kinase A via a phosphatidylinositol 4,5-bisphosphate-dependent mechanism. Proceedings of the National Academy of Sciences USA 96: 5820–5825.

Liu J-H, Bijlenga P, Fischer-Lougheed J, Occhiodoro T, Kaelin A, Bader CR, Bernheim L (1998). Role of an inward rectifier K^+ current and of hyperpolarization in human myoblast formation. Journal of Physiology 510: 467–476.

Liu Y, Holmgren M, Jurman ME, Yellen G (1997). Gated access to the pore of a voltage-dependent K^+ channel. Neuron 19: 175–184.

Liu Y, Liu D, Heath L, Meyers DM, Krafte DS, Wagoner PK, Silvia CP, Yu W, Curran ME (2001). Direct activation of an inwardly rectifying potassium channel by arachadonic acid. Molecular pharmacology 59: 1061–1068.

Loew A, Ho Y-K, Blundell T, Bax B (1998) Phoducin induces a structural change in transducin βγ. Structure 6: 1007–1019.

Logothetis DE, Kim D, Northup JK, Neer EJ, Clapham DE (1988) Specificity of action of guanine nucleotide-binding regulatory protein subunits on the cardiac muscarinic K^+ channel. Proceedings of the National Academy of Sciences USA 85:5814–5818

Logothetis DE, Kurachi Y, Galper J, Neer EJ, Clapham DE (1987) The βγ subunits of GTP-binding proteins activate the muscarinic K^+ channel in heart. Nature 325:321–326

Lopatin AN, Makhina EN, Nichols CG (1994). Potassium channel block by cytoplasmic polyamines as the mechanism of intrinsic rectification. Nature 372: 366–369

Lopatin AN, Makhina EN, Nichols CG (1995). The mechanism of inward rectification of potassium channels: 'long-pore plugging' by cytoplasmic polyamines. Journal of General Physiology 106: 923–955.

Lopatin AN, Nichols CG (1996a) $[K^+]$ dependence of polyamine-induced rectification in inward rectifier potassium channels (IRK1, Kir2.1). Journal of General Physiology 108: 105–113.

Lopatin AN, Nichols CG (1996b) $[K^+]$ dependence of open channel conductance in cloned inward rectifier potassium channels (IRK1, Kir2.1). Biophysical Journal 71: 682–694

Lopez-Barneo J, Hoshi T, Heinemann SH, Aldrich RW (1993). Effects of external cations and mutations in the pore region on C-type inactivation of *Shaker* potassium channels. Receptors & Channels 1:61–71.

Loussouarn G, Makhina EN, Rose T, Nichols CG (2000). Structure and dynamics of the pore of inwardly rectifying K_{ATP} channels. Journal of Biological Chemistry 275: 1137–1144.

Lü Q, Miller, C (1995). Silver as a probe of pore forming residues in a potassium channel. Science 268: 304–307.

Lu T, Nguyen B, Zhang X, Yang J (1999a). Architecture of a K^+ channel inner pore revealed by stoichiometric covalent modification. Neuron 22: 571–580.

Lu T, Zhu Y-G, Yang J (1999b). Cytoplasmic amino and carboxyl domains form a wide intracellular vestibule in an inwardly rectifying potassium channel. Proceedings of the National Academy of Sciences USA 96: 9926–9931.

Lu Z, Klem AM, Ramu Y (2001). Ion conduction pore is conserved among potassium channels. Nature 413: 809–813.

Lu Z, MacKinnon R (1994) Electrostatic tuning of Mg^{2+} affinity in an inward-rectifier K^+ channel. Nature 371: 243–246.

Lu Z, MacKinnon R (1997). Purification, characterization, and synthesis of an inward-rectifier K^+ channel inhibitor from scorpion toxin. Biochemistry 36: 6936–6940.

Lucas JJ, Mellström B, Colado MI, Naranjo JR (1993). Molecular mechanisms of pain: $serotonin_{1A}$ receptor agonists trigger transactivation by *c-fos* of the prodynorphin gene in spinal cord neurons. Neuron 10: 599–611.

Luchian T, Dascal N, Dessauer C, Platzer D, Davidson N, Lester HA, Schreibmayer W (1997) A C-terminal peptide of the GIRK1 subunit directly blocks the G protein-activated K^+ channel (GIRK) expressed in *Xenopus* oocytes. Journal of Physiology 505:13–22

Lüscher C, Jan LY, Stoffel M, Malenka RC, and Nicoll RA (1997) G protein-coupled inwardly rectifying K^+ channels (GIRKs) mediate postsynaptic but not presynaptic transmitter actions in hippocampal neurons. Neuron 19: 687–695.

Ma D, Zerangue N, Lin YF, Collins A, Yu M, Jan YN, Jan LY (2001). Role of ER export signals in controlling surface potassium channel numbers. Science 291: 316–319.

MacKinnon R, Cohen SL, Kuo A, Lee A, Chait BT (1998). Structural conservation in prokaryotic and eukaryotic potassium channels. Science 280: 106–109.

MacKinnon R, Yellen G (1990). Mutations affecting TEA blockade and ion permeation in voltage-activated K^+ channels. Science 250: 276–279.

Madison DV, Lancaster B, Nicoll RA (1987) Voltage clamp analysis of cholinergic action in the hippocampus.Journal of Neuroscience 7:733–741

Makhina EN, Kelley AJ, Lopatin AN, Nichols CG (1994). Cloning and expression of a novel human brain inward rectifier potassium channel. Journal of Biological Chemistry 269: 20468

Mark MD, Herlitze S (2000) G-protein mediated gating of inward-rectifier K^+ channels. European Journal of Biochemistry 267: 5830–5836.

Matsuda H (1988). Open-state substructure of inwardly rectifying potassium channels revealed by magnesium block in guinea-pig heart cells. Journal of Physiology 397: 237–258.

Matsuda H, Matsuura H, Noma A (1989). Triple-barrel structure of inwardly rectifying K^+ channels revealed by Cs^+ and Rb^+ block in guinea-pig heart cells. Journal of Physiology 413: 139–157.

Matsuda H, Saigusa A, Irisawa H (1987). Ohmic conductance through the inwardly rectifying K channel and blocking by internal Mg^{2+}. Nature 325: 156–159.

Matsuda H, Stanfield PR (1989). Single inwardly rectifying potassium channels in cultured muscle cells from rat and mouse. Journal of Physiology 414:111–124

McCarron JG, Halpern W (1990). Potassium dilates rat cerebral arteries by two independent mechanisms. American Journal of Physiology 259: H902–H908.

McGurk JF, Bennett MVL, Zukin RS (1990). Polyamines potentiate responses of N-methyl-D-aspartate receptors expressed in Xenopus oocytes. Proceedings of the National Academy of Sciences USA 87: 9971–9974.

McNicholas CM, Guggino WB. Schweibert EM, Hebert SC, Giebisch G, Egan ME (1996). Sensitivity of a renal K^+ channel (ROMK2) to the inhibitory sulfonylurea compound glibenclamide is enhanced by coexpression with the ATP-binding cassette transporter cystic fibrosis transmembrane regulator. Proceedings of the National Academy of Sciences USA 93: 8083–8088.

Medina I, Krapivinsky G, Arnold S, Kovoor P, Krapivinsky L, Clapham DE (2000) A switch mechanism for Gβγ activation of I_{KACh}. Journal of Biological Chemistry 275: 29709–29716.

Meves H (1994). Modulation of ion channels by arachidonic acid. Progress in Neurobiology 43: 175–186.

Meyer T, Wellner-Kienitz, M-C, Biewald A, Bender K, Eickel A, Pott L (2001) Depletion of phosphatidylinositol 4,5-bisphosphate by activation of phospholipase C-coupled receptors causes slow inhibition but not desensitization of G protein-gated inward rectifier K^+ current in atrial myocytes. Journal of Biological Chemistry 276: 5650–5658.

Mi H, Deerinck TJ, Jones M, Ellisman MH, Schwarz TL (1996). Inwardly rectifying K^+ channels that may participate in K^+ buffering are localized in microvilli of Schwann cells. Journal of Neuroscience 16: 2421–2429.

Mihara S, North RA, Surprenant A (1987) Somatostatin increases an inwardly rectifying potassium conductance in guinea-pig submucous plexus neurones. Journal of Physiology 390:335–355

Minor DL, Masseling SJ, Jan YN, Jan LY (1999) Transmembrane structure of an inwardly rectifying potassium channel. Cell 96: 879–891.

Miyake M, Christie MJ, North RA (1989) Single potassium channels opened by opioids in rat locus ceruleus neurons. Proceedings of the National Academy of Sciences USA 86:3419–3422

Miyazawa A, Fujiyoshi Y, Stowell M, Unwin N (1999). Nicotinic acetylcholine receptor at 4.6Å resolution: transverse tunnels in the channel wall. Journal of Molecular Biology 288: 765–786.

Morishige K-I, Takahashi CN, Jahangir A, Yamada CM, Koyama H, Zanelli JS, Kurachi Y (1994) Molecular cloning and functional expression of a novel brain-specific inward rectifier potassium channel. FEBS Letters 346: 251–256.

Morishita R, Nakayama H, Isobe T, Matsuda T, Hashimoto Y, Okano T, Fukada Y, Mizuno K, Ohno S, Kozawa O, Kato K, Asano T (1995) Primary structure of a γ subunit of G protein, γ_{12}, and its phosphorylation by protein kinase C. Journal of Biological Chemistry 270:29469–29475

Müllner C., Vorobiov D, Bera AK, Uezono Y, Yakubovich D, Frohnwieser-Steinecker B, Dascal N, Schreibmayer W (2000) Journal of General Physiology 115: 547–557.

Mumby SM, Casey PJ, Gilman AG, Gutowski S, Sternweis PC (1990) G protein γ subunits contain a 20-carbon isoprenoid. Proceedings of the National Academy of Sciences USA 87:5873–5877

Muntz KH, Sternweis PC, Gilman AG, Mumby SM (1992) Influence of γ subunit prenylation on association of guanine nucleotide-binding regulatory proteins with membranes. Molecular Biological Cell 3: 49–61

Murer G, Adelbrecht C, Lauritzen I, Lesage F, Lazdunski M, Agid Y, Raisman-Vozari R (1997) An immunocytochemical study on the distribution of two G-protein-gated inward rectifier potassium channels (GIRK2 and GIRK4) in the adult rat brain. Neuroscience 80:345-357

Nakajima S, Iwasaki S, Obata K (1962) Delayed rectification and anomalous rectification in frog's skeletal muscle membrane. Journal of General Physiology 46:97–115

Nakajima S, Nakajima Y, Peachey LD (1973). Speed of repolarization and morphology of glycerol-treated frog muscle fibres. Journal of Phyisology 234: 465–480.

Nakajima S, Takahashi K (1966) Post-tetanic hyperpolarization and electrogenic Na pump in stretch receptor neurone of crayfish. Journal of Phyisology 187:105–127

Nakajima Y, Nakajima S (1994) Signal transduction mechanisms of tachykinin effects on ion channels. In: Buck SH (ed). The tachykinin receptors. Humana Press Inc. Totowa, New Jersey, pp 285–327

Nakajima Y, Nakajima S, Inoue M (1988) Pertussis toxin-insensitive G protein mediates substance P-induced inhibition of potassium channels in brain neurons. Proceedings of the National Academy of Sciences USA 85:3643–3647

Nakajima Y, Nakajima S, Kozasa T (1996) Activation of G protein-coupled inward rectifier K^+ channels in brain neurons requires association of G protein $\beta\gamma$ subunits with cell membrane. FEBS Letters 390:217–220

Nakamura N, Suzuki Y, Sakuta H, Ookata K, Kawahara K, Hirose S (1999). Inwardly rectifying K^+ channel Kir7.1 is highly expressed in thyroid follicular cells, intestinal epithelial cells and choroid plexus epithelial cells: Implication for a functional coupling with Na^+,K^+-ATPase. Biochemical Journal 342: 329–336.

Nakamura TY, Artman M, Rudy B, Coetzee WA (1998). Inhibition of rat ventricular IK1 with antisense oigonucleotides targeted to Kir2.1 mRNA. American Journal of Physiology 274: H892–900.

Namba N, Inagaki N, Gonoi T, SeinoY, Seino S (1996). Kir2.2v: a possible negative regulator of the inwardly rectifying K+ channel Kir2.2. FEBS Letters 386:211–214.

Navarro B, Kennedy ME, Velimirovic B, Bhat D, Peterson AS, Clapham DE (1996). Nonselective and G-insensitive *weaver* K^+ channels. Science 272: 1950–1953.

Nehring RB. Wischmeyer E, Doring F, Veh RW, Sheng M, Karschin A (2000). Neuronal inwardly rectifying K^+ channels differentially couple to PDZ proteins of the PSD-95/SAP90 family. Journal of Neuroscience 20: 156–162.

Nelson CS, Marino JL, Allen CN (1997). Cloning and characterization of Kir3.1 (GIRK1) C-terminal alternative splice variants. Molecular Brain Research 46: 185–196.

Neubig RR (1994) Membrane organization in G-protein mechanisms. FASEB Journal 8:939–946

Newman EA (1984). Regional specialization of retinal glial cell membrane. Nature 309: 155–157.
Newman EA (1993) Inward-rectifying potassium channels in retinal glial (Müller) cells. Journal of Neuroscience 13:3333–3345
Nichols CG, Lopatin AN (1997) Inward rectifier potassium channels. Annual Reviews of Physiology 59:171–191
Nilius B, Schwarz G, Droogmans G (1993) Modulation by histamine of an inwardly rectifying potassium channel in human endothelial cells. Journal of Physiology 472:359–371
Niu XW, Meech RW (1998). The effect of polyamines on K(ATP) channels in guinea-pig ventricular myocytes. Journal of Physiology 508: 401–411.
Noma A (1983) ATP-regulated K^+ channels in cardiac muscle. Nature 305:147–148
North RA (1989) Drug receptors and the inhibition of nerve cells. British Journal of Pharmacology 98:13–28
North RA, Uchimura N (1989) 5-hydroxytryptamine acts at 5-HT_2 receptors to decrease potassium conductance in rat nucleus accumbens neurones. Journal of Physiology 417:1–12
North RA, Williams JT (1985) On the potassium conductance increased by opioids in rat locus coeruleus neurones. Journal of Physiology 364:265–280
North RA, Williams JT, Surprenant A, Christie MJ (1987) μ and δ receptors belong to a family of receptors that are coupled to potassium channels. Proceedings of the National Academy of Sciences USA 84:5487–5491
Oh U, Ho Y-K, Kim D (1995) Modulation of the serotonin-activated K^+ channel by G protein subunits and nucleotides in rat hippocampal neurons. Journal of Membrane Biological 147:241–253
Ohmori H (1978). Inactivation kinetics and steady-state current noise in the anomalous rectifier of tunicate egg cell membranes. Journal of Physiology 281: 77–89.
Oishi K, Omori K, Ohyama H, Shingu K, Matsuda H (1998). Neutralization of aspartate residues in the murine inwardly rectifying K^+ channel IRK1 affects the substate behaviour in Mg^{2+} block. Journal of Physiology 510: 675–683.
Olesen SP, Bundgaard M (1993) ATP-dependent closure and reactivation of inward rectifier K^+ channels in endothelial cells. Circulation Research 73:492–495
Olesen SP, Davies PF, Clapham DE (1988). Muscarinic-activated K^+ current in bovine aortic endothelial cells. Circulation Research 62: 1059–1064.
Oliva C, Cohen IS, Pennefather P (1990). The mechanism of rectification of $i(K_1)$ in canine Purkinje myocytes. Journal of General Physiology 96: 299–318.
Orkand RK, Nicholls JG, Kuffler SW (1966) Effect of nerve impulses on the membrane potential of glial cells in the central nervous system of amphibia. Journal of Neurophysiology 29:788–806
Osmanovic SS, Shefner SA (1987) Anomalous rectification in rat locus coeruleus neurons. Brain Research 417:161–166
Papazian DM, Schwarz TL, Tempel BL, Jan YN, Jan LY (1987) Cloning of genomic and complementary DNA from shaker, a putative potassium channel gene from Drosophila. Science 237:749–753
Pardo LA, Heinemann SH, Terlau H, Ludewig U, Lorra C, Pongs O, Stühmer W (1992). Extracellular K^+ specifically modulates a rat brain K^+ channel. Proceedings of the National Academy of Sciences USA 89: 2466–2470.
Pascual JM, Shieh C-C, Kirsch GE, Brown AM (1995). K^+ pore structure revealed by reporter cysteines at inner and outer surfaces. Neuron 14: 1055–1063.
Patil N, Cox DR, Bhat D, Faham M, Myers RM, Peterson AS (1995). A potassium channel mutation in weaver mice implicates membrane excitability in granule cell differentiation. Nature Genetics 11: 126–129.

Paulson OB, Newman EA (1987). Does the release of potassium from astrocyte endfeet regulate cerebral blood flow? Science 237: 896–898.
Pawson T, Scott JD (1997) Signaling through scaffold, anchoring, and adaptor proteins. Science 278: 2075–2080.
Pearson WL, Dourado M, Schreiber M, Salkoff L, Nichols CG. (1999). Expression of a functional Kir4 family inward rectifier K^+ channel from a gene cloned from mouse liver. Journal of Physiology 514: 655–665.
Pearson WL, Nichols CG (1998). Block of the Kir2.1 channel pore by alkylamine analogues of endogenous polyamines. Journal of General Physiology 112: 351–363.
Peng L, Zhang H, Hirsch J, Logothetis DE (2000) The yeast $\beta\gamma$ subunits of G proteins inhibit GIRK4 channels. Biophysical Journal 78: 465A.
Penington NJ, Kelly JS, Fox AP (1993a) Whole-cell recordings of inwardly rectifying K^+ currents activated by 5-HT_{1A} receptors on dorsal raphe neurones of the adult rat. Journal of Physiology (Lond) 469:387–405
Penington NJ, Kelly JS, Fox AP (1993b) Unitary properties of potassium channels activated by 5-HT in acutely isolated rat dorsal raphe neurones. Journal of Physiology 469:407–426
Pennefather P, Oliva C, & Mulrine N (1992) Origin of the potassium and voltage dependence of the cardiac inwardly rectifying K-current, I (K_1). Biophysical Journal 61: 448–462.
Pennefather PS, Heisler S, MacDonald JF (1988) A potassium conductance contributes to the action of somatostatin-14 to suppress ACTH secretion. Brain Research 444:346–350
Périer F, Radeke CM, Vandenberg CA (1994) Primary structure and characterization of a small-conductance inwardly rectifying potassium channel from human hippocampus. Proceedings of the National Academy of Sciences USA 91: 6240–6244.
Perillán PR, Li X, Potts EA, Chen M, Bredt DS, Simard JM (2000). Inward rectifier K^+channel Kir2.3 (IRK3) in reactive astrocytes from adult rat brain. Glia 31: 181–192.
Perozo E, Cortes DM, Cuello LG (1999). Structural rearangements underlying K^+-channel activation gating. Science 285: 73–78.
Pessia M, Tucker SJ, Lee K, Bond CT, Adelman JP (1996). Subunit positional effects revealed by novel heteromeric inwardly rectifying K^+ channels. EMBO Journal 15: 2980–2987.
Peters R (1981) Translational diffusion in the plasma membrane of single cells as studied by fluorescence microphotolysis. Cell Biology International Reports 5:733–760
Pfaffinger PJ, Martin JM, Hunter DD, Nathanson NM, Hille B (1985) GTP-binding proteins couple cardiac muscarinic receptors to a K channel. Nature 317:536–538
Plaster NM, Tawil R, Tristani-Firouzi M, Canun S, Bendahhou S, Tsunoda A, Donaldson MR, Iannaccone ST, Brunt E, Barohn R, Clark J, Deymeer F, George Jr AL, Fish FA, Hahn A, Nitu A, Ozdemir C, Serdaroglu P, Subramony SH, Wolfe G, Fu YH, Ptacek LJ (2001). Mutations in Kir2.1 cause the developmental and episodic electrical phenotypes of Andersen's syndrome. Cell 105: 511–519.
Posner BA, Mukhopadhyay, S, Tesmer JJ, Gilman AG, Ross EM (1999) Modulation of the affinity and selectivity of RGS protein interaction with G_α subunits by a conserved aspargine/serine residue. Biochemistry 38: 7773–7779.
Prior HM, Webster N, Quinn K, Beech DJ, Yates MS (1998). K^+-induced dilation of a small renal artery: no role for inward rectifier K^+ channels. Cardiovascular Research 37: 780–790.

Pronin AN, Gautam N (1992) Interaction between G-protein β and γ subunit types is selective. Proceedings of the National Academy of Sciences USA 89:6220–6224

Qu Z, Yang Z, Cui N, Zhu G, Liu C, Xu H, Chanchevalap S, Shen W, Wu J, Li Y, Jiang C (2000). Gating of inward rectifier K^+ channels by proton-mediated interactions of N- and C-terminal domains. Journal of Biological Chemistry 275: 31573–31580.

Qu Z, Zhu G, Yang Z, Cui N, Li Y, Chanchevalap S, Sulaiman S, Haynie H, Jiang C (1999). Identification of a critical motif responsible for gating of Kir2.3 channel by intracellular protons. Journal of Biological Chemistry 274: 13783–13789.

Quayle JM, Dart C, Standen NB (1996). The properties and distribution of inward rectifier potassium currents in pig coronary arterial smooth muscle. Journal of Physiology 494: 715–726.

Raab-Graham KF, Radeke CM, Vandenberg CA (1994) Molecular cloning and expression of a human heart inward rectifier potassium channel. Neuroreport 5:2501–2501.

Raab-Graham KF, Vandenberg CA (1998). Tetrameric subunit structure of the native brain inwardly rectifying potassium channel K(ir)2.2. Journal of Biological Chemistry 273:19699–19707.

Rakowski RF, Gadsby DC, De Weer P (1989) Stoichiometry and voltage dependence of the sodium pump in voltage-clamped, internally dialyzed squid giant axon. Journal of General Physiology 93:903–941

Rang HP, Ritchie JM (1968) On the electrogenic sodium pump in mammalian non-myelinated nerve fibres and its activation by various external cations. Journal of Physiology 196: 183–221.

Rettig J, Heinemann SH, Wunder F, Lorra C, Parcej DN, Dolly JO, Pongs O (1994). Inactivation properties of voltage gated K^+ channels altered by the presence of a β subunit. Nature 369: 289–294.

Reuveny E, Jan YN, Jan LY (1996). Contributions of a negatively charged residue in the hydrophobic domain of the IRK1 inwardly rectifying K^+ channel to K^+-selective permeation. Biophysical Journal 70: 754–761.

Reuveny E, Slesinger PA, Inglese J, Morales JM, Iñiguez-Lluhi JA, Lefkowitz RJ, Bourne HR, Jan YN, Jan LY (1994) Activation of the cloned muscarinic potassium channel by G protein βγ subunits. Nature 370:143–146

Rohács T, Chen J, Prestwich GD, Logothetis DE (1999). Distinct specificities of inwardly rectifying K^+ channels for phosphoinositides. Journal of Biological Chemistry 274: 36065–36072.

Ross, E.M. and Wilkie, T.M. (2000) GTPase-activating proteins for heterotrimeric G proteins: regulators of G protein signaling (RGS) and RGS-like proteins. Annual Reviews of Biochemistery **69:** 795–827.

Roy ML, Sontheimer H (1995) β-Adrenergic modulation of glial inwardly rectifying potassium channels. Journal of Neurochemistery 64:1576–1584

Roychowdhury S, Rasenick MM (1997) G protein β1γ2 subunits promote microtubule assmebly. Journal of Biological Chemistery 272:31576–31581

Roychowdhury S, Wang N, Rasenick M (1993) G protein binding and G protein activation by nucleotide transfer involve distinct domains on tubulin: regulation of signal transduction by cytokeletal elements. Biochemistry 32:4955–4961

Ruppersberg JP, Fakler B (1996). Complexity of the regulation of K(ir)2.1 K^+ channels. Neuropharmacology 35: 887–893.

Sabirov RZ, Okada Y, Oiki S (1997a). Two-sided action of protons on an inward rectifier K^+ channel (IRK1). Pflügers Archiv 433: 428–434.

Sabirov RZ, Tominaga T, Miwa A, Okada Y, Oiki S (1997b). A conserved arginine residue in the pore region of an inward rectifier K channel (IRK1) as an external barrier for cationic blockers. Journal of General Physiology 110: 665–677.

Sadja R, Smadja K, Alagem N, Reuveny E (2001). Coupling Gβγ-dependent activation to channel opening via pore elements in inwardly rectifying potassium channels. Neuron 29: 669–680.

Saitoh O, Kubo Y, Miyatani Y, Asano T, Nakata H (1997) RGS8 accelerates G-protein-mediated modulation of K^+ currents. Nature 390:525–529

Sakmann B, Noma A, Trautwein W (1983) Acetylcholine activation of single muscarinic K^+ channels in isolated pacemaker cells of the mammalian heart. Nature 303:250–253

Sakmann B, Patlak J, Neher E (1980) Single acetylcholine-activated channels show burst-kinetics in presence of desensitizing concentrations of agonist. Nature 286:71–73.

Schmidt CJ, Thomas TC, Levine MA, Neer EJ (1992) Specificity of G protein β and γ subunit interactions. Journal of Biological Chemistery 267:13807–13810

Schreibmayer W, Dessauer CW, Vorobiov D, Gilman AG, Lester HA, Davidson N, Dascal N (1996) Inhibition of an inwardly rectifying K^+ channel by G-protein α-subunits. Nature 380:624–627

Schulte U, Hahn H, Konrad M, Jeck N, Derst C, Wild K, Weidemann S, Ruppersberg JP, Fakler B, Ludwig J (1999). pH gating of ROMK (K(ir)1.1) channels: control by an Arg-Lys-Arg triad disrupted in Bartter syndrome. Proceedings of the National Academy of Sciences USA 96: 15298–15303.

Schulte U, Weidemann S, Ludwig J, Ruppersberg JP, Fakler B (2001). K^+-dependent gating of $K_{ir}2.1$ channels is linked to pH gating through a conformational change in the pore. Journal of Physiology 534:59–70.

Sharon D, Vorobiov D, Dascal N (1997) Positive and negative coupling of the metabotropic glutamate receptors to a G protein-activated K^+ channel, GIRK, in *Xenopus* oocytes. Journal of General Physiology 109:477–490

Shelton PA, Davies NW, Conley EC, Sutcliffe, MJ & Stanfield PR (1995). Effects of N- and C-terminal deletions from the murine inward rectifier potassium channel IRK1. Japanese Journal of Physiology 45 (Suppl 2): S110.

Shieh R-C (2000). Mechanisms for the time-dependent decay of inward currents through cloned Kir2,1 channels expressed in *Xenopus* ococytes. Journal of Physiology 526: 241–252.

Shieh R-C, Chang J-C, Arreola J (1998). Interaction of Ba^{2+} with the pores of the cloned inward rectifier K^+ channels Kir2.1 expressed in *Xenopus* oocytes. Biophysical Journal 75: 2313–2322.

Shieh R-C, Chang J-C, Kuo C-C (1999). K^+ binding sites and interactions between permeating K^+ ions at the external pore mouth of an inward rectifier K^+ channel (Kir2.1). Journal of Biological Chemistry 274: 17424–17430.

Shieh R-C, John SA, Lee J-K, Weiss JN (1996). Inward rectification of the IRK1 channel expressed in Xenopus oocytes: Effects of intracellular pH reveal an intrinsic gating mechanism. Journal of Physiology 494: 363–376.

Shieh R-C, Lee J-K (2001). Ammonium ions induce inactivation of Kir2.1 potassium channels expressed in Xenopus oocytes Journal of Physiology 535: 359-370.

Shin KS, Park JY, Kwon H, Chung CH, Kang MS (1997). A possible role of inwardly rectifying K^+ channels in chick myoblast differentiation. American Journal of Physiology 272: C894–900.

Shuck ME, Bock JH, Benjamin CW, Tsai TD, Lee KS, Slightom JL, Bienkowski MJ (1994a) Cloning and characterization of multiple forms of the human kidney ROM-K potassium channel. Journal of Biological Chemistery 269:24261–24270

Shuck ME, Piser TM, Bock JH, Slightom JL, Lee KS, Bienkowski MJ (1994b) Cloning and characterization of two K^+ inward rectifier (K(ir) 1.1) potassium channel homologs from human kidney (K(ir)1.2 and K(ir) 1.3) Journal of Biological Chemistry 272:586–593.

Shyng S-L, Barbieri A, Gumusboga A, Cukras C, Pike L, Davis JN, Stahl PD, Nichols CG (2000) Modulation of nucleotide sensitivity of ATP-sensitive potassium channels by phophatidylinositol-4-phosphate 5-kinase. Proceedings of the National Academy of Sciences USA 97: 937–941.

Shyng S-L, Nichols CG (1997). Octameric stoichiometry of the K(ATP) channel complex. Journal of General Physiology 110: 655–664.

Shyng S-L, Nichols CG (1998) Membrane phospholipid control of nucleotide sensitivity of K_{ATP} channels. Science 282: 1138–1144

Shyng S-L, Sha Q, Ferrigni T, Lopatin AN & Nichols CG (1996). Depletion of intracellular polyamines relieves inward rectification of potassium channels. Proceedings of the National Academy of Sciences USA 93: 12014–12019.

Sigworth FJ (1985) Open channel noise. I. Noise in acetylcholine receptor currents suggests conformational fluctuation. Biophysical Journal 47: 709–720.

Silver MR, DeCoursey TE (1990). Intrinsic gating of inward rectifier in bovine pulmonary artery endothelial cells in the presence or absence of internal Mg^{2+}. Journal of General Physiology 96: 109–133.

Silverman SK, Lester HA, Dougherty DA (1996) Subunit stoichiometry of a heteromultimeric G protein-coupled inward-rectifier K^+ channel. Journal of Biological Chemistery 271:30524–30528

Silverman SK, Lester HA, Dougherty DA (1998). Asymmetrical contributions of subunit pore regions to ion selectivity in an inward rectifier K^+ channel. Biophysical Journal 75: 1330–1339.

Simonds WF, Butrynski JE, Gautam N, Unson CG, Spiegel AM (1991) G-protein βγ dimers. Journal of Biological Chemistery 266:5363–5366

Sine SM, Claudio T, Sigworth FJ (1990) Activation of *Torpedo* acetylcholine receptors expressed in mouse fibroblasts. Journal of General Physiology 96:395–437

Sine SM, Steinbach JH (1984) Activation of a nicotinic acetylcholine receptor. Biophysical Journal 45:175–85

Slesinger P, Patil N, Liao YJ, Jan YN, Jan LY, Cox DR (1996). Functional effects of the mouse *weaver* mutation on G-protein-gated inwardly rectifying K^+ channels. Neuron 16, 321–331.

Slesinger PA, Reuveny E, Jan YN, Jan LY (1995) Identification of structural elements involved in G protein gating of the GIRK1 potassium channel. Neuron 15:1145–1156

Snow, B.E., Betts, L., Mangion, J., Sondek, J., and Siderovski, D.P. (1999) Fidelity of G protein β-subunit association by the G protein γ-subunit-like domains of RGS6, RGS7, and RGS11. Proceedings of the National Academy of Sciences USA **96**: 6489–6494.

Snow, B.E., Krumins, A.M., Brothers, G.M., Lee, S.-F., Wall, M.A., Chung, S., Mangion, J., Arya, S., Gilman, A.G., and Siderovski, D.P. (1998) A G protein γ subunit-like domain shared between RGS11 and other RGS proteins specifies binding to $G_{\beta5}$ subunits. Proceedings of the National Academy of Sciences USA **95**: 13307–13312.

So I, Ashmole I, Davies NW, Sutcliffe MJ, Stanfield PR (2001). The K^+ channel signature sequence of murine Kir2.1: mutations that affect microscopic gating but not ionic selectivity. Journal of Physiology 531: 37–50.

Sodickson DL, Bean BP (1996) $GABA_B$ receptor-activated inwardly rectifying potassium current in dissociated hippocampal CA3 neurons. Journal of Neuroscience 16: 6374–6385.

Soejima M, Noma A (1984) Mode of regulation of the ACh-sensitive K-channel by the muscarinic receptor in rabbit atrial cells. Pflügers Archiv 400:424–431

Sokolova O, Kolmakova-Partensky L, Grigorieff N (2001). Three-dimensional structure of a voltage-gated potassium channel at 2.5nm resolution. Structure 9: 215–220.

Sondek J, Bohm A, Lambright DG, Hamm HE, Sigler PB (1996) Crystal structure of a G_A protein $\beta\gamma$ dimer at 2.1A resolution. Nature 379:369–374

Soom M, Schonherr R, Kubo Y, Kirsch C, Klinger R, Heinemann SH (2001). Multiple PIP_2 binding sites in Kir2.1 inwardly rectifying potassium channels. FEBS Letters 490: 49–53.

Sorota S, Tsuji Y, Tajima T, Pappano AJ (1985) Pertussis toxin treatment blocks hyperpolarization by muscarinic agonists in chick atrium. Circulation Research 57:748–758

Spalding BC, Senyk O, Swift JG, Horowicz P (1981). Unidirectional flux ratio for potassium ions in depolarized frog skeletal muscle. American Journal of Physiology 10: C68–C75.

Spassova M, Lu Z (1998). Coupled ion movement underlies rectification in an inward-rectifier K^+ channel. Journal of General Physiology 112: 211–221.

Spauschus A, Lentes K-U, Wischmeyer E, Dibmann E, Karschin C, Karschin A (1996) A G-protein-activated inwardly rectifying K^+ channel (GIRK4) from human hippocampus associates with other GIRK channels. Journal of Neuroscience 16:930–938

Stampe P, Arreola J, Perez-Cornejo P, Begenisich T (1998). Non independent K^+ movement through the pore in IRK1 potassium channels. Journal of General Physiology 112: 475–484.

Stampe P, Begenisich T (1996). Unidirectional K^+ fluxes through recombinant Shaker potassium channels expressed in single Xenopus oocytes. Journal of General Physiology 107: 449–457.

Standen NB, Stanfield PR (1978a). A potential- and time-dependent blockade of inward rectification in frog skeletal muscle fibres by barium and strontium ions. Journal of Physiology 280: 169–191.

Standen NB, Stanfield PR (1978b). Inward rectification in skeletal muscle: a blocking particle model. Pflügers Archiv 378: 173–176.

Standen NB, Stanfield PR (1979). Potassium depletion and sodium block of potassium currents under hyperpolarization in frog sartorius. Journal of Physiology 294: 497–520.

Standen NB, Stanfield PR (1980). Rubidium block and rubidium permeability of the inward rectifier of frog skeletal muscle fibres. Journal of Physiology 304: 415–435.

Stanfield PR, Ashcroft FM, Plant TD (1981). Gating of a muscle potassium channel and its dependence on the permeating ion species. Nature 289: 509–510.

Stanfield PR, Davies NW, Shelton PA Khan IA, Brammar WJ, Standen NB, Conley EC (1994a) The intrinsic gating of inward rectifier K^+ channels expressed from the murine IRK1 gene depends on voltage, K^+ and Mg^{2+}. Journal of Physiology 475: 1–7.

Stanfield PR, Davies NW, Shelton PA Sutcliffe MJ Khan IA, Brammar WJ, Conley EC (1994b) A single aspartate residue is involved in both intrinsic gating and blockage by Mg^{2+} of the inward rectifier, IRK1. Journal of Physiology 478: 1–6.

Stanfield PR, Nakajima Y, Yamaguchi K (1985) Substance P raises neuronal membrane excitability by reducing inward rectification. Nature 315:498–501

Stevens EB, Shah BS, Pinnock RD, Lee K (1999) Bombesin receptors inhibit G protein-coupled inwardly rectifying K^+ channels expressed in Xenopus oocytes through a protein kinase C-dependent pathway. Molecular Pharmacology 55: 1020–1027.

Stockklausner C, Ludwig J, Ruppersberg JP, Klöcker N (2001). A sequence motif responsible for ER export and surface expression of Kir2.0 inward rectifier K^+ channels. FEBS Letters 493: 129–133.

Stonehouse AH, Pringle JH, Norman RI, Stanfield PR, Conley EC, Brammar WJ (1999). Characterisation of Kir2.0 proteins in the rat cerebellum and hippocampus by polyclonal antibodies. Histochemistry & Cell Biology 112: 457–465.

Sui JL, Chan KW, Logothetis DE (1996) Na^+ activation of the muscarinic K^+ channel by a G-protein-independent mechanism. J Gen Physiol 108:381–391

Sui JL, Petit-Jacues J, Logothetis DE (1998) Activation of the atrial K_{ACh} channel by the $\beta\gamma$ subunits of G proteins or intracellular Na^+ ions depends on the presence of phophatidylinositol phosphates. Proceedings of the National Academy of Sciences USA 95: 1307–1312.

Surprenant A, North RA (1988) Mechanism of synaptic inhibition by noradrenaline acting at α_2-adrenoceptors. Proceedings of the Royal Society B234:85–114

Sutton KG, Dolphin AC, Scott RH (1993). Inhibition of voltage-activated Ca^{2+} currents from cultured sensory neurones by spermine, argiotoxin-636 and a synthetic arginine polyamine. Molecular Neuropharmacology 3:37–43.

Taglialatela M, Ficker E, Wible BA, Brown AM (1995). C-terminus determinants for Mg^{2+} and polyamine block of the inward rectifier K^+ channel IRK1. EMBO Journal 14: 5532–5541.

Taglialatela M, Wible BA, Caporaso R, Brown AM (1994). Specification of pore properties by the carboxyl terminus of inwardly rectifying K^+ channels. Science 264: 844–847.

Takahashi N, Morishige K-I, Jahangir A, Yamada M, Findlay I, Koyama H, Kurachi Y (1994) Molecular cloning and functional expression of cDNA encoding a second class of inward rectifier potassium channels in the mouse brain. Journal of Biological Chemistry 269: 23274–23279.

Takano K, Asano S, Yamashita N (1994) Activation of G protein-coupled K^+ channels by dopamine in human GH-producing cells. American Journal of Physiology 266: E318–E325

Takano K, Stanfield PR, Nakajima S, Nakajima Y (1995) Protein kinase C-mediated inhibition of an inward rectifier potassium channel by substance P in nucleus basalis neurons. Neuron 14:999–1008

Takano K, Yasufuku-Takano J, Kozasa T, Nakajima S, Nakajima Y (1997) Different G proteins mediate somatostatin-induced inward rectifier K+ currents in murine brain and endocrine cells. Journal of Physiology 502:559–567

Takano K, Yasufuku-Takano J, Kozasa T, Singer WD, Nakajima S, Nakajima Y (1996) $G_{q/11}$ and PLC-β1 mediate substance P-induced inhibition of an inward rectifier K^+ channel in brain neurons. Journal of Neurophysiology 76:2131–2136

Takao K, Yoshii M, Kanda A, Kokubun S, Nukada T (1994) A region of the muscarinic-gated atrial K^+ channel critical for activation by G protein $\beta\gamma$ subunits. Neuron 13:747–755

Takumi T, Ishii T, Horio Y, Morishige K-I, Takahashi N, Yamada M, Yamashita T, Kiyama H, Sohmiya K, Nakahishi S, Kurachi Y (1995). A novel ATP-dependent inward rectifier potassium channels expressed predominantly in glial cells. Journal of Biological Chemistry 270: 16339–16346.

Tang W, Yang X-C. (1994). Cloning a novel human brain inward rectifier potassium channel and its functional expression in *Xenopus* oocytes. FEBS Letters 348: 239–243

Tempel BL, Papazian DM, Schwarz TL, Jan YN, Jan LY (1987) Sequence of a probable potassium channel component encoded at shaker locus of drosophila. Science 237:770–775

Thomas RC (1972) Electrogenic sodium pump in nerve and muscel cells. Physiological Reviews 52:563–594

Thompson GA, Leyland MJ, Ashmole I, Sutcliffe MJ, Stanfield PR (2000). Residues beyond the selectivity filter of the K^+ channel Kir2.1 regulate permeation and block by external Rb^+ and Cs^+. Journal of Physiology 526: 231–240.

Tinker A, Jan YN. Jan,LY (1996). Regions responsible for the assembly of inwardly rectifying potassium channels. Cell 87: 857–868.

Tong Y, Brandt GS, Li M, Shapovalov G, Slimko E, Karschin A, Dougherty DA, Lester HA (2001). Tyrosine decaging leads to substantial membrane trafficking during modulation of an inward rectifier potassium channel. Journal of General Physiology 117: 103–118.

Töpert C, Döring F, Wischmeyer E, Karschin C, Brockhaus J, Ballanyi K, Derst C, Karschin A (1998). Kir2.4: A novel K^+ inward rectifier channel associated with motoneurons of cranial nerve nuclei. Journal of Neuroscience 18: 4096–4105.

Trautwein W, Dudel J (1958) Zum Mechanismus der Membranwirkung des Acetylcholin an der Herzmuskelfaser. Pflügers Archiv 266:324–334

Trussell LO, Jackson MB (1985) Adenosine-activated potassium conductance in cultured striatal neurons. Proceedings of the National Academy of Sciences USA 82:4857–4861

Trussell LO, Jackson MB (1987) Dependence of an adenosine-activated potassium current on a GTP-binding protein in mammalian central neurons. Journal of Neuroscience 7:3306–3316

Tsaur M-L, Menzel S, Lai F-P, Espinosa III R, Concannon P, Spielman RS, Hanis CL, Cox NJ, Le Beau MM, German MS, Jan LY, Bell GI, Stoffel M (1995) Isolation of a cDNA clone encoding a K(ATP) channel-like protein expressed in insulin-secreting cells, localization of the human gene to chromosome band 21q22.1, and linkage studies with NIDDM. Diabetes 44: 592–596.

Tucker SJ, Gribble FM, Proks P, Trapp S, Ryder TJ, Haug T, Reimann F, Ashcroft FM (1998). Molecular determinants of K_{ATP} channel inhibition by ATP. EMBO Journal 17: 3290–3296.

Tucker SJ, Pessia M, Adelman JP (1996) Muscarine-gated K^+ channel: subunit stoichiometry and structural domains essential for G protein stimulation. American Journal of Physiology 271:H379–H385

Uchimura N, North RA (1990) Muscarine reduces inwardly rectifying potassium conductance in rat nucleus accumbens neurones.Journal of Physiology 422:369–380

Vandenberg CA (1987) Inward rectification of a potassium channel in cardiac ventricular cells depends on internal magnesium ions. Proceedings of the National Academy USA 84: 2560–2564.

Velimirovic BM, Gordon EA, Lim NF, Navarro B, Clapham DE (1996) The K^+ channel inward rectifier subunits form a channel similar to neuronal G protein-gated K^+ channel. FEBS Letters 379:31–37

Velimirovic, B., Koyano, K, Nakajima, S., and Nakajima, Y. (1991) Substance P, somatostatin and met-enkephalin regulate the same K- channel in cultured noradrenergic neurons from the locus coeruleus. Society for Neuroscience Abstracts, **17**: 1474.

Velimirovic BM, Koyano K, Nakajima S, Nakajima Y (1995) Opposing mechanisms of regulation of a G-protein-coupled inward rectifier K^+ channel in rat brain neurons. Proceedings of the National Academy of Sciences USA 92:1590–1594

Wall MA, Coleman DE, Lee E, Iñiguez-Lluhi JA, Posner BA, Gilman AG, Sprang SR (1995) The structure of the G protein heterotrimer $G_{i\alpha1}\beta_1\gamma_2$. Cell 83:1047–1058

Wallinga W, Meijer SL, Alberink MJ, Vliek M, Wienk ED, Ypey DL (1999) Modelling action potentials and membrane currents of mammalian skeletal muscle fibres in coherence with potassium concentration changes in the T-tubular system. European Biophysics Journal 28: 317–329.

Wang HS, McKinnon D (1996) Modulation of inwardly rectifying currents in rat sympathetic neurones by muscarinic receptors.Journal of Physiology 492.2:467–478

Wang W, Giebisch G (1991) Dual modulation of renal ATP-sensitive K^+ channel by protein kinases A and C. Proceedings of the National Academy of Sciences USA 88:9722–9725

Warmke JW, Ganetsky B (1994). A family of potassium channel genes related to *eag* in *Drosophila* and mammals. Proceedings of the National Academy USA 91: 3438–3442.

Watson S, Arkinstall S (1994) The G-protein linked receptor facts book Academic Press, London. 427pp.

Wei J, Hodes ME, Piva R, Feng Y, Wang Y, Ghetti B, Dlouhy SR (1998). Characterization of murine Girk2 transcript isoforms: structure and differential expression. Genomics 51: 379–390.

Weidmann S (1951). Effect of current flow on the membrane potential of cardiac muscle. Journal of Physiology 115: 227–236.

Weiger T, Hermann A (1994). Polyamines block Ca^{2+}-activated K^+ channels in pituitary tumor cells (GH3). Journal of Membrane Biology 140: 133–142.

Wells JA (1990). Additivity of mutational effects of proteins. Biochemistry 29: 8509–8517.

Wessel R, Kristan Jr WB, Kleinfeld D (1999). Supralinear summation of synaptic inputs by an invertegbrate neuron: dendritic gain is mediated by an 'inward rectifier' K^+ current. Journal of Neuroscience 19: 5875–5888.

Wible BA, Taglialatela M, Ficker E, Brown AM (1994). Gating of inwardly rectfying K^+ channels localized to a single negatively charged residue. Nature 371: 246–249.

Wickman K, Clapham DE (1995) Ion channel regulation by G proteins. Physiologycal Reviews 75:865–885

Wickman KD, Iñiguez-Lluhi JA, Davenport PA, Taussig R, Krapivinsky GB, Linder ME, Gilman AG, Clapham DE (1994) Recombinant G-protein βγ-subunits activate the muscarinic-gated atrial potassium channel. Nature 368:255–257

Wickman K, Nemec J, Gendler SJ, Clapham DE (1998) Abnormal heart rate regulation in *GIRK4* knockout mice. Neuron 20: 103–114.

Williams JT, Colmers WF, Pan ZZ (1988a) Voltage-and ligand-activated inwardly rectifying currents in dorsal raphe neurons in vitro. Journal of Neuroscience 8:3499–3506

Williams JT, Egan TM, North RA (1982) Enkephalin opens potassium channels on mammalian central neurones. Nature 299: 74–77

Williams JT, North RA, Tokimasa T (1988b) Inward rectification of resting and opiate-activated potassium currents in rat locus coeruleus neurons. Journal of Neurosci ence 8:4299–4306

Williams K (1997) Modulation and block of ion channels: a new biology of polyamines. Cellular Signalling 9: 1–13.

Wilson GF, Chiu SY. (1990) Ion channels in axon and Schwann cell membranes at paranodes of mammalian myelinated fibers studied with patch clamp. Journal of Neuroscience 10: 3263–3274

Wimpey TL, Chavkin C (1991) Opioids activate both an inward rectifier and a novel voltage-gated potassiu conductance in the hippocampal formation. Neuron 6:281–289

Wischmeyer E, Döring F, Karschin A (1998). Acute suppression of inwardly rectifying Kir2.1 channels by direct tyrosine kinase phosphorylation. Journal of Biological Chemistry 273: 34063–34068.

Wischmeyer E, Döring F, Wischmeyer E, Spauschus A, Thomzig A, Veh R, Karschin A (1997) Subunit interactions in the assembly of neuronal Kir3.0 inwardly rectifying K^+ channels. Molecular and Cellular Neuroscience 9: 194–206.

Wischmeyer E, Karschin A (1996). Receptor stimulation causes slow inhibition of IRK1 inwardly rectifying K+ channels by direct protein kinase A-mediated phosphorylation. Proceedings of the National Academy of Sciences USA 93: 5819–5823.

Womack KB, Gordon SE, He F, Wensel TG, Lu C-C, Hilgemann DW (2000). Do phosphatidylinositides modulate vertebrate phototransduction? Journal of Neuroscience 20: 2792–2799.

Wu D, Jiang H, Katz A, and Simon MI (1993) Identification of critical regions on phospholipase C-β1 required for activation of G-proteins. Journal of Biological Chemistry 268: 3704–3709.

Xie L-H, Horie M, Takano M (1999) Phospholipase C-linked receptors regulate the ATP-sensitive potassium channel by means of phosphatidylinositol 4, 5-bisphosphate metabolism. Proceedings of the National Academy of Sciences USA 96:15292–15297.

Yakubovich D, Pastushenko V, Bitler A, Dessauer CW, Dascal N (2000) Slow modal gating of single G protein-activated K^+ channels expressed in *Xenopus* oocytes. Journal of Physiology 524:737–755.

Yamada M, Inanobe A, Kurachi Y (1998) G protein regulation of potassium ion channels. Pharmacological Reviews 50: 723–757.

Yamada M, Kurachi Y (1995) Spermine gates inward-rectifying muscarinic but not ATP-sensitive K^+ channels in rabbit atrial myocytes. Journal of Biological Chemistry 270: 9289–9294.

Yamashita N, Shibuya N, Ogata E (1988) Requirement of GTP on somatostatin-induced K^+ current in human pituitary tumor cells. Proceedings of the National Academy of Sciences USA 85:4924–4928

Yang J, Jan YN, Jan LY (1995a). Control of rectification and permeation by residues in two distinct domains in an inward rectifier K^+ channel. Neuron 14: 1047–1054.

Yang J, Jan YN, Jan LY (1995b). Determination of the subunit stoichiometry of an inwardly rectifying potassium channel. Neuron 15: 1441–1447.

Yang J, Yu M, Jan YM, Jan LY (1997). Stabilisation of ion selectivity filter by pore loop ion pairs in an inwardly rectifying potassium channel. Proceedings of the National Academy USA 94:1568–1572.

Yang Z, Xu H, Cui N, Qu Z, Chanchevalap S, Shen W, Jiang C (2000). Biophysical and molecular mechanisms underlying the modulation of heteromeric Kir4.1–Kir5.1 channels by CO_2 and pH. Journal of General Physiology 116: 33–45.

Yano H, Philipson LH, Kugler JL, Tokuyama Y, Davis EM, Le Beau MM, Nelson DJ, Bell GI, Takeda J (1994) Alternative splicing of human inwardly rectifying K^+ channel ROMK1 mRNA. Molecular Pharmacology 45:854–860

Yasuda H, Lindorfer MA, Woodfork KA, Fletcher JE, Garrison JC (1996) Role of the prenyl group on the G protein γ subunit in coupling trimeric G proteins to A1 adenosine receptors. Journal of Biological Chemistry 271:18588–18595

Yatani A, Codina J, Brown AM, Birnbaumer L (1987) Direct activation of mammalian atrial muscarinic potassium channels by GTP regulatory protein G_K. Science 235: 207–211.

Yi BA, Lin Y-F, Jan YN, Jan LY (2001). Yeast screen for constitutively active mutant G protein-activated potassium channels. Neuron 29: 657–667.

Zaritsky JJ, Eckman DM, Wellman GC, Nelson MT, Schwarz TL (2000). Targeted disruption of Kir2.1 and Kir2.2 genes reveals the essential role of the inwardly rectifying K^+ current in K^+-mediated vasodilation. Circulation Research 87: 160–166.

Zaritsky JJ, Redell JB, Tempel BL Schwarz TL (2001). The consequences of disrupting cardiac inwardly rectifying K^+ current (I_{K1}) as revealed by the targeted deletion of the murine *Kir2.1* and *Kir2.2* genes. Journal of Physiology 533: 697–710.

Zerangue N, Schwappach B, Jan YN, Jan LY (1999). A new ER trafficking signal regulates the subunit stoichiometry of plasma membrane K_{ATP} channels. Journal of Physiology 22: 537–548.

Zhainazarov AB, Ache BW (1999). Effects of phosphatidylinositol 4,5-bisphosphate and phosphatidylinositol 4-phosphate on a Na^+-gated nonselective cation channel. Journal of Neuroscience 19: 2929–2937.

Zhang H, He C, Yan X, Mirshahi T, Logothetis DE (1999) Activation of inwardly rectifying K^+ channels by distinct PtdIns(4,5)P_2 interactions. Nature Cell Biology 1: 183–188

Zhang, J.H. and Simonds, W.F. (2000). Copurification of brain G-protein b5 with RGS6 and RGS7. Journal of Neuroscience **20:** RC59, 1–5.

Zhou H, Chepilko S, Schutt W, Choe H, Palmer LG, Sackin H (1996). Mutations in the pore region of ROMK enhance Ba^{2+} block. American Journal of Physiology 271: C1949–C1956.

Zhou H, Tate SS, Palmer LG (1994) Primary structure and functional properties of an epithelial K channel. American Journal of Physiology 266:C809–C824

Zhou M, Morais-Cabral JH, Mann S, MacKinnon R (2001). Potassium channel receptor site for the inactivation gate and quaternary amine inhibitors. Nature 411:657–661.

Zhu G, Chanchevalap S, Cui N, Jiang C (1999a). Effects of intra- and extracellular acidifications on single channel Kir2.3 currents. Journal of Physiology 516: 699–710.

Zhu G, Liu C, Qu Z, Chanchevalap S, Xu H, Jiang C (2000). CO_2 inhibits specific inward rectifier K^+ channels by decreases in intra- and extracellular pH. Journal of Biological Chemistry 274: 11643–11646.

Zhu G, Qu Z, Cui N, Jiang C (1999b). Suppression of Kir2.3 activity by protein kinase C phosphorylation of the channel protein at threonine 53. Journal of Biological Chemistry 274: 11643–11646.

Zhu L, Wu X, Wu MB, Chan KW, Logothetis DE, Thornhill WB (2001). Cloning and characterization of G protein-gated inward rectifier K^+ channel (GIRK1) isoforms from heart and brain. Journal of Molecular Neuroscience 16: 21–32.

Chloroplast Quest: A Journey From the Cytosol into the Chloroplast and Beyond

Ute C. Vothknecht, J. Soll

Botanisches Institut der Ludwig-Maximilian-Universität München, Menzinger Str. 67, D-80638 München, Germany

Contents

1 Summary 183
2 Introduction 183
  2.1 Evolutionary Origin of Chloroplasts 183
    2.1.1 Primary Endosymbiosis 183
    2.1.2 Secondary Endosymbiosis 187
  2.2 Chloroplast-Nuclear Communication 188
3 Chloroplast Targeting 189
  3.1 Targeting by Presequence 189
    3.1.1 Structure of Chloroplast Transit Sequences 190
    3.1.2 Transit Sequence Prediction 190
    3.1.3 Lipid Bilayer Interaction 191
  3.2. Targeting without Presequence 192
    3.2.1 Insertion into the Outer Envelope Membrane 192
4 The General Import Pathway 193
  4.1 Cytosolic Factors and the Guidance Complex 195
  4.2 The Toc Complex 196
    4.2.1 Toc159 196
    4.2.2 Toc34 197
    4.2.3 Toc75 198
    4.2.4 Toc64 199
  4.3 The Intermembrane Space 199
  4.4 The Tic Complex 199
    4.4.1 Tic110 200
    4.4.2 Tic62 200
    4.4.3 Tic55 201
    4.4.4 Tic40 201
    4.4.5 Tic22/Tic20 201
  4.5 Stromal Factors 202
  4.6 Protein Import in *Arabidopsis*, a Genomic Approach 203

Reviews of Physiology, Biochemistry, and Pharmacology, Vol. 145

5 Intraplastidal Sorting 204
5.1 Insertion into the Inner Envelope Membrane 204
5.2 Thylakoid Targeting 205
5.2.1 The Sec-pathway 205
5.2.2 The ΔpH-dependent or TAT-pathway 206
5.2.3 The SRP pathway 207
5.2.4 Spontaneous Insertion 208
5.3 Intraplastidal Vesicle Transport 209
6 Protein Translocation into Complex Plastids 209
6.1 Bipartite Presequences 210
6.2 Cytosol to Plastid via ER 210
7 Evolution of the Import Apparatus 211
8 Outlook 214
Acknowledgements 214
Abbreviations 214
References 215

1 Summary

Chloroplasts are characteristic organelles of plants and algae and the site of oxygenic photosynthesis. They are surrounded by a double membrane and possess an internal membrane system, the thylakoids, on which the photosynthetic machinery is located. They originated more than 1.2 billion years ago from an endosymbiotic event between an already photosynthetic ancestor of present day cyanobacteria and a mitochondriate host cell. During the transformation of the internalized cyanobacterium into a cell organelle most of the genetic information of the endosymbiot got lost or was transferred into the nucleus of the host. Chloroplast proteins encoded by nuclear genes are synthesized on cytoplasmic ribosomes and have to be relocated into the organelle. This is achieved by a proteinaceous import machinery in the outer and inner envelope of the chloroplasts. Proteins destined for the thylakoid membrane and the thylakoid lumen are further translocated by several different pathways into or across this membrane. The subject of this review is the quest of nuclear encoded chloroplast proteins into the organelle and to their final suborganellar location.

2 Introduction

2.1 Evolutionary Origin of Chloroplasts

2.1.1 Primary Endosymbiosis

Chloroplasts originated from an endosymbiotic event between an ancestor of today's cyanobacteria and a host cell that had previously acquired a mitochondrion (Mereschkowsky 1905; Margulis 1970). The mitochondriate host cell internalized the prokaryotic cell leading to a close symbiosis of the two. Subsequent events lead to the transformation of the independent organism into a semi-autonomous cell organelle (Fig. 1). It is believed that this so-called primary endosymbiosis occurred only once in the history of plant evolution and gave rise to all plastids (Palmer 2000). Thus, present day chloroplasts share a common ancestry, they are monophyletic. The cyanobacterial endosymbiot had already acquired the capacity for oxygenic photosynthesis, i.e., the two photosystems, the oxygen evolving system and a specialized internal membrane system, the thylakoids (Mereskowsky 1905). This capacity was maintained in the emerging cell organelle, the chloroplast. Other features

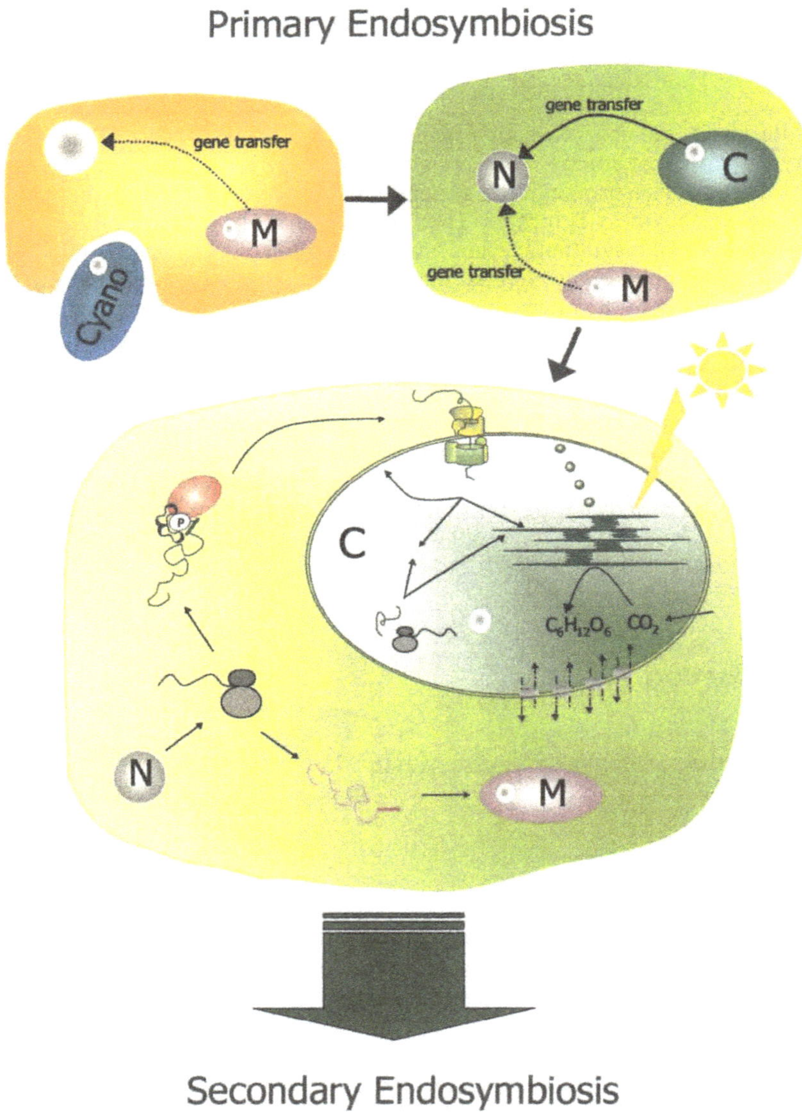

Fig. 1. Evolution of chloroplasts and complex plastids by primary and secondary endosymbiosis. All chloroplasts originated from a singular endosymbiotic event by which an ancestor of today's cyanobacteria was internalized by a mitochondriate host and transformed into a cell organelle. Complex plastids originated from multiple endosymbiotic events between photoautotrophic eukaryotes and photoheterotrophic host cells.

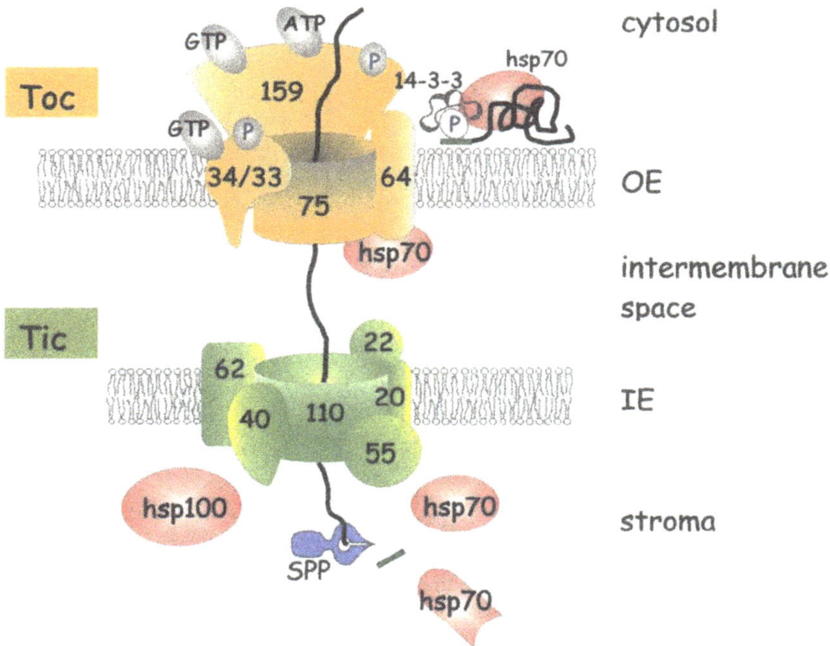

Fig. 2. Import into chloroplasts via the general import pathway. Chloroplast targeted proteins are synthesized with an N-terminal cleavable transit sequence that can be phophorylated by a cytosolic kinase. The precursor protein is then recognized by a cytosolic guidance complex and directed to the chloroplast surface. Alternatively precursor protein can travel to the import machinery without the help from the guidance complex. Proteins are translocated simultaneously through both envelope membranes by a proteinaceous import machinery called Toc (Translocon on the outer envelope of chloroplasts) and Tic (Translocon on the inner envelope of chloroplasts). Translocation requires ATP and is regulated by phosphorylation and GTP. It additionally needs chaperones in the intermembrane space and the stroma. The transit peptide is cleaved by a soluble Stromal Processing Peptidase (SPP).

that were retained during chloroplast formation include fatty acid biosynthesis, nitrite reduction, and amino acid biosynthesis. Chloroplasts still contain multiple copies of their own circular genome and they have the competence for DNA replication, transcription, and translation. Nevertheless, during the formation of the chloroplast, the majority of the genetic information of the original endosymbiot was transferred to the nucleus of the host cell or was lost completely (Martin and Herrmann 1998).

Modern chloroplasts encode only about 50 to 150 of the estimated 2000 to 5000 proteins that make up the plastidal proteome. For many or even all of

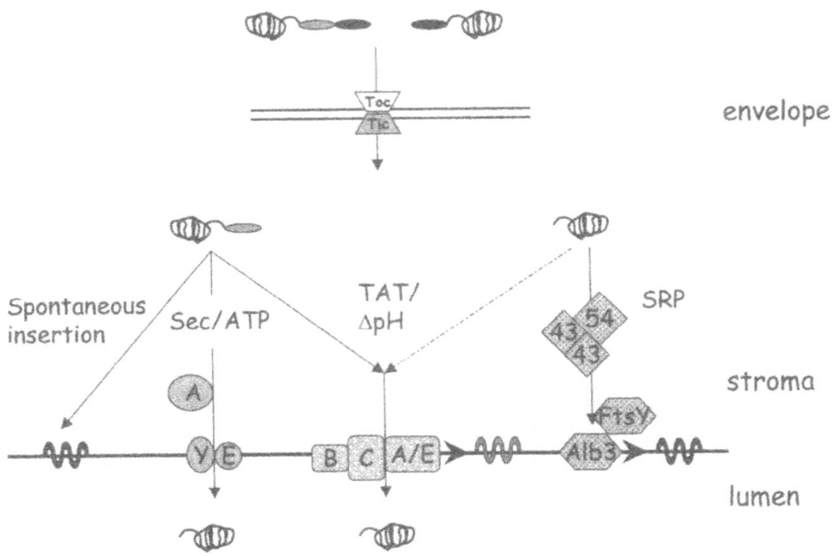

Fig. 3. Pathways for protein translocation into and across the thylakoid membrane. At least four different pathways exist for thylakoid translocation: Sec-, TAT-, SRP-pathway, and spontaneous insertion. All of these are homologous to known translocation pathways in bacteria and/or eukaryotic cells. Three of these pathways utilize an additional N-terminal signal sequence. Evidence exists that some of the components might in fact be shared between different pathways.

their functions chloroplasts rely on components encoded in the nucleus. Due to its endosymbiotic origin the chloroplast (like the mitochondrion) is surrounded by a double membrane. It is believed that these membranes derived from the inner and outer membrane of the cyanobacterial symbiot (Cavalier-Smith 1987). Proteins that are synthesized in the cytosol have to translocate across both of these membranes in order to enter the chloroplast (Fig. 1+2). Thus, in concert with the formation of the chloroplast and the relocation of the majority of its genome to the nucleus, the need for plastidal targeting and import machinery emerged. The import machinery as we know it today was not inherited in toto from the cyanobacterium but was build up anew around a few preexisting components. The newly created import machinery was further connected to complex regulatory networks that gave the nucleus the command over the organelle. A monophyletic origin of all plastids also implies that the import machinery was developed once and should thus be conserved between species, giving room mainly for adaptation (Cavalier-Smith 2000). The mechanisms employed in the insertion of proteins into or transfer over the thylakoid membrane are derivatives of pathways which already existed in the cyanobacterial ancestor. They utilize translocation machineries that are also present in non-photosynthetic bacteria (Fig. 3).

Plastids derived from this primary endosymbiosis can nowadays be found in green algae and land plants, red algae and glaucophytes. The latter retained the bacterial cell wall in addition to the two envelopes and thus might be the earliest branch of plastid evolution (Palmer 2000).

2.1.2 Secondary Endosymbiosis

The evolution of organism did not stop at the primary endosymbiotic event that created chloroplasts. After the establishment of early phototrophic eukaryotic cells (both red and green algae), several incidents occurred in which eukaryotic host cells internalized these phototrophic eukaryotes (Gibbs 1978). This happened multiple times and secondary endosymbiosis gave rise to a number of different lineages. In all of these cases most of the features of the engulfed eukaryote were lost over time and in the end little more than the chloroplast remained from the eukaryotic symbiot. Nevertheless, a shade of the original eukaryotic cell persisted. The organelles created by secondary endosymbiosis are surrounded by one or even two membranes in addition to the outer and inner envelope of the original chloroplast. They are hence called complex plastids. In some but not all cases a residual nucleus of the eukaryotic symbiot is still found in the periplastidal space between the outer membrane of the complex plastid and the envelope of the original organelle. It is called the nucleomorph and contains DNA that sets it apart from the eukaryotic host cell of the endosymbiosis. The presence of eukaryotic-type 80S ribosomes implies a continuous eukaryotic type protein synthesis in the periplastidic compartment (Hansmann and Eschbach 1990; McFadden et al. 1994).

Complex plastids can be distinguished by their descent from red or green algae, their number of membranes, and the presence and absence of a nucleomorph. The origin from red algae is more abundant and can be found in heterokonts, cryptomonads, apicomplexa and dinoflagellates. With the exception of dinoflagellates all of these organisms contain complex plastids with four membranes, but only the cryptomonades do still possess a nucleomorph. Green algae gave rise to the complex plastids found in chlorarachniophytes (four membrane, nucleomorph) and *Euglena* (three membranes, without nucleomorph).

Some of the lineages containing complex plastids are quite unexpected, i.e. a group of non-photosynthetic parasites, called apicomplexa (McFadden et al. 1996; Köhler et al. 1997). These include important human and animal pathogens such as *Plasmodium spp.* (malaria causing agent) and *Toxoplasma gondii*. It can be imagined that the apicomplexa once lived as independent, photoautotrophic organism but their photosynthetic function has been lost subsequent to or together with their adaptation to a parasitic life style. Nevertheless, the residual plastids obviously retain an indispensable function for these organisms, e.g. fatty acid and isoprenoid biosynthesis (McFadden et al. 1996). The plastids offer a good target for specific drug development. These drugs are targeted on components specific to an organelle that does not exist in humans or animals infected by these parasites. This has been successfully exploited, e.g. by the use of such antibiotics as doxycycline, lincosamides and macrolides (Fichera and Roos, 1997; McFadden and Roos, 1999). Also new drugs directed to fight malaria have originated from the unique target offered by these organelles.

2.2 Chloroplast-Nuclear Communication

During the evolution of chloroplasts most of the genetic information of the original endosymbiotic cell got lost or was transferred into the nucleus of the host (Martin and Herrmann 1998). This gene transfer resulted in a necessary communication between the organelle and the nucleus to ensure the transcription and translation of nuclear-coded chloroplast proteins in concert with transcription and translation in the organelle and its biochemical needs (Goldschmidt-Clermont 1998; Rodermel 2001). This requirement for communication is greatly enhanced by the fragmentation of plastidal multisubunit protein complexes and metabolic pathways. They contain proteins encoded both in the organelle and the nuclear genome (Herrmann 1999; Jarvis 2001; Rodermel 2001). Assembly of these complexes requires an ordered expression of all components, since the failure of assembling multiprotein complexes often results in a degradation of the unassembled proteins. The complexity of coordination between the nucleus and the organelles is increased by the fact that nuclear-coded proteins are expressed from genes that exist in a single or very few copies. In comparison, plastid DNA is present in several hundred copies in the chloroplast with a single cell having up to 100 or more chloroplasts at a given time (Bendich 1987).

For a long time the signals passing between the organelle and the nucleus remained enigmatic. Yet lately, great progress has been made towards our understanding of chloroplast-nuclear communication. Nuclear-encoded factors that regulate organelle proteins in a post-translational manner have been described. They are used to force a nuclear control on the organelle gene expression (Rochaix 2001). It has also been clear for a long time that this control is not unidirectional but that chloroplasts signal their developmental and/or metabolic status back to the nucleus and directly control nuclear gene expression. The nature of this so-called plastidal-factor is just beginning to emerge. Instead of a single factor, multiple elements seem to be involved in this retrograde signaling. They are believed to include light, porphyrins, reactive oxygen intermediates, and carotenoids. The involvement of redox signaling in this communication is interesting since there is growing indication for a redox regulation of plastidal import (Caliebe et al. 1997; Hirohashi et al. 2001; Küchler et al. submitted). It is easy to imagine that import of proteins is one of the control points of nucleus-organelle interaction.

3 Chloroplast Targeting

3.1 Targeting by Presequence

Most nuclear-encoded chloroplast proteins are synthesized on cytosolic ribosomes with an N-terminal cleavable presequence (Dobberstein et al. 1977). This presequence is called the transit peptide and it is cleaved inside the chloroplast to release the mature protein. It is believed that the transit peptide is all that is required to direct a protein into the chloroplast. Yet there is no common motif or sequence pattern identified that characterizes typical transit sequences for chloroplast targeting. Plastidal transit sequences are very variable in length, ranging from 20 to more than 120 residues. They also display an immense variation in their primary amino acid sequence (von Heijne et al. 1989; Schein et al. 2001). Nevertheless, they all fulfill a common objective. They all target proteins to the chloroplast and across the chloroplast double membrane. They also prevent any mistargeting of plastidal proteins into other organelles, e.g. mitochondria, at least within homologous systems (de Castra Silva Filho 1996). Dual targeting, however is sometimes employed by organisms to target the same protein to multiple cellular locations. This is most often achieved by alternative splicing or multiple start codons that result in proteins with different N-termini. Yet there are also known examples of proteins targeted into both mitochondria and chloroplasts by an identical transit sequence *in vivo* (reviewed in Small et al. 1998).

Regarding our present knowledge on the way transit sequences promote translocation into chloroplasts one should keep in mind that most experiments have been carried out using the transit peptides of only a small set of plastidal proteins, most often either pFdI (precursor protein of ferredoxin isoprotein 1) or pSSU (precursor protein of the small subunit of ribulose-bisphosphate carboxylase/oxygenase). Both precursor proteins are easily obtained by in vitro transcription/translation and are willing accomplices of *in vitro* chloroplast import experiments. Further studies are necessary to show that the information obtained for these and other precursor proteins are valid for the whole set of chloroplast imported proteins.

3.1.1
Structure of Chloroplast Transit Sequences

Despite their variability in length and primary amino acid sequence, certain characteristics are shared by all chloroplast transit sequences. In general, the N-proximal part of the transit sequence contains predominantly positively charged residues as well as the amino acids glycine and proline. A central domain of variable length is enriched in hydroxylated amino acids such as serine and threonine and is deficient in acidic residues. The C-proximal part most closely to the cleavage site and the start of the mature protein even has a vague conserved consensus sequence (von Heinje et al. 1989; Bartling et al. 1990).

Mitochondrial presequences are characterized by their ability to form a distinct secondary structure. This, rather than the primary sequence is the basis for recognition by the mitochondrial import machinery. So far no such conserved secondary structure could be identified for chloroplast targeting sequences. Thus neither primary nor secondary amino acid sequence appears to be the foremost property that guides chloroplast import. An important feature specific to chloroplast transit sequences is a phosphorylation site. Phosphorylation of major chloroplast preproteins by a cytosolic kinase of unknown nature has been shown exclusively for chloroplast import (Waegemann and Soll 1996). And while a great deal is known about how certain changes in the transit sequence affect the recognition, import and processing of proteins, the exact means by which the translocation machinery recognizes a chloroplast transit peptide remains elusive.

3.1.2
Transit Sequence Prediction

Ever since the finding that signal sequences target proteins to their appropriate location within the cell, a great effort was made to predict the potential target organelle of a protein by its deduced presequence. In case of chloroplast transit sequences, due to their great variability and lack of consensus in amino acid sequence and structure, this task has proved rather difficult. Several neural network models have been created and made available for researchers to help elucidating organellar localization. These programs use different algorithms and they can predict a chloroplast localization with different accuracy. Their algorithms are based on the sequence information of known chloroplast-targeted proteins. Yet, even with the genome sequence of *Arabidopsis* being completed, very few of the predicted 2000 to 5000 chloroplast proteins are detected to date. Peltier et al (2000) systematically identified thylakoid membrane proteins from *Arabidopsis* isolated by two-dimensional gel electrophoresis and mass spectroscopic analysis. These proteins were tested against the prediction programs. Extrapolating from their results they estimated that a genome wide analysis of organelle-targeted proteins would create a great number of false positives by missing out many of the correct genes at the same time. This and other proteomic approaches, like the systematic identification of lumenal proteins by Schubert et al. (2001), should help to further increase our knowledge about chloroplast targeting sequences and thus make their prediction more accurate. In the end they will enable us to finally identified most of the chloroplast-targeted proteins including their transit sequences at least in *Arabidopsis*.

3.1.3 Lipid Bilayer Interaction

Major progress has been made recently in elucidating and understanding the composition and function of proteinaceous components involved in chloroplast import. Much less studies have focused on the question whether and how the lipid bilayer of the chloroplast envelope plays a vital part in the import process. Both, the outer and the inner envelope of chloroplasts have a unique lipid composition (Bruce 1998). They contain several unusual lipids, including monogalactosyldiacylglyceride (MGDG), digalactosyldiacylglyceride (DGDG), sulfolipid, and phosphatidylglycerol (PG). The outer envelope (OE) is the only membrane exposed to the cytosol that contains MGDG and DGDG. Lipids have thus been implicated in plastidal import and it has been speculated that the transit sequence interacts specifically with the lipid bilayer of the chloroplast prior to engaging the import machinery (van't Hof and de Kruijff 1995; Bruce 1998). In vitro experiments with signal sequences for the endoplasmic reticulum (ER), bacterial membranes, mitochondria and chloroplasts, have shown that they are capable of an interaction with lipid bilayers in the absence of membrane proteins. The overall positive net charge of most chloroplast transit sequences would allow such interaction with the negatively charged chloroplast surface. A specific interaction has been shown in vitro between the transit peptides of pSSU and pFdI and liposomes containing chloroplast specific lipids such as PG and MGDG (van't Hof et al. 1993; Pilon et al. 1995; Pinnaduwage and Bruce 1996). Kerber and Soll (1992) furthermore confirmed that an alteration of the lipid composition of the outer envelope by phosholipase C treatment affected the import properties of the chloroplasts. These data are supported by in vivo studies on *dgd1*, an *Arabidopsis* DGDG-deficient mutant, that displays a reduced efficiency in chloroplast import (Chen and Li 1998). A possible interaction of precursor protein with outer envelope membrane lipids might also solve the question concerning recognition of chloroplast transit sequences. It cannot be excluded that lipid interaction will induce a specific secondary structure of the transit peptide that is then recognized by the import machinery.

3.2 Targeting without Presequence

Nuclear-encoded chloroplast proteins that are translated with a cleavable transit peptide enter the chloroplast by the general import pathway (see below). In contrast, a subset of chloroplast-targeted proteins do not contain transit peptides but are synthesized in the cytosol in their mature form (Salomon et al. 1990; Li et al. 1991; Ko and Ko 1992; Fischer et al. 1994; Kessler et al. 1994). So far this group includes nearly all integral membrane proteins of the outer envelope (OEP, outer envelope protein). Yet there is growing evidence that also internal chloroplast proteins translocate into the chloroplast without a cleavable targeting sequence. This is still a point of debate and future research will have to show whether such proteins indeed exist and by what pathway they translocate into the chloroplast.

3.2.1 Insertion into the Outer Envelope Membrane

Very little is known about how OEPs are targeted to the chloroplast and how their insertion into the outer envelope membrane is mediated. Two different pathways could be envisioned for the translocation of these proteins. They could insert into the membrane without the assistant of a proteinaceous insertion machinery in a process dubbed spontaneous insertion. This process might or might not include a specific interaction with the lipids of the outer envelope as discussed for proteins with a transit sequence (van't Hof and de Kruijff 1995; Bruce 1998). Alternatively, targeting to and insertion into the outer envelope could be assisted by factors in the cytosol or the envelope membrane itself.

Original experiments using protease treated isolated chloroplasts suggested that the insertion is indeed independent of any proteinaceous components and ATP and it was referred to as the "OM14" pathway (Salomon et al. 1990; Li et al. 1991; Fischer et al. 1994; Li and Chen 1996). Owing to recent studies on several different outer envelope proteins a more meticulous and diverse picture is emerging. At least two outer envelope proteins, OEP14 and COM70, contain specific targeting signals inside their mature protein sequence (Wu and Ko 1993; Li and Chen 1996). The OEP14 signal resembles targeting sequences for the ER, yet it is specific for chloroplasts. Tu and Li (2000) showed that insertion of OEP14 requires a component sensitive to trypsin and N-ethylmaleimide. They could also show that OEP14 competes with other outer envelope proteins for insertion. These experiments indicate that a proteinaceous machinery rather than spontaneous insertion is the means by which these OEPs insert into the chloroplast membrane. Other, opposing experiments showed that peaOEP14, OEP7 from spinach, OEP21, and also DGDG synthase 1 do not require any thermolysin-sensitive components or

ATP (Salomon et al. 1990; Li et al. 1991; Froehlich et al. 2001). Indeed OEP7 inserts spontaneously into protein free liposomes, indicating that no peptide cofactors are essential. Rather the topology of OEP7 is influenced by the lipid asymmetry of the chloroplast envelope membranes (Schleiff et al. 2001). The import of OEP24 on the other hand was shown to be thermolysin-resistent but was stimulated by ATP (Fischer et al. 1994). Import of pea Toc34, an integral component of the import machinery in pea chloroplasts, is stimulated by both, a thermolysin-sensitive component and ATP (Tsai et al. 1999). However, Toc34 can also insert spontaneously into liposomes (Schleiff et al. submitted), indicating a multiple step process with distinct requirements for each step. Other recent experiments by Hiltbrunner and coworkers (2001) suggested that atToc33 (the *Arabidopsis* homologue of peaToc34) functions in the insertion of atToc160, another integral component of the Toc machinery in the outer envelope. This corroborates earlier data by Muckel and Soll (1996) who showed that Toc159 (the pea homologue of atToc160) insertion into the outer envelope requires a protease-sensitive component at the chloroplast surface. Further experimental data are necessary to elucidate whether these proteins use different, identical or partially overlapping insertion pathways. Also the exact nature of these pathway(s) has to be revealed. Furthermore it can not be excluded that at least in certain cases the insertion of an outer envelope protein into the membrane occurs unassisted while the assembly of the protein into a functional complex requires proteinaceous components and other factors.

One notable exception is the insertion of Toc75, an integral component of the import machinery of chloroplasts. Toc75 contains a cleavable bipartite presequence and it is targeted to the outer envelope via the general import pathway (Schnell et al. 1994; Tranel et al. 1995). The N-proximal transit sequence is cleaved in the stroma before Toc75 is inserted into the outer envelope and the C-proximal presequence is processed by an unknown mechanism (Tranel et al. 1995). Toc75 does not enter into the stroma completely but engages the import machinery of both the outer and the inner envelope just deep enough to have its target sequence cleaved. This unusual insertion of Toc75 is interesting in view of its evolutionary origin (see 7).

4
The General Import Pathway

Once a precursor protein has made contact with the chloroplast surface and the envelope import machinery it can be translocated into the organelle. Most proteins containing a cleavable transit sequence are believed to engage a general import pathway. This pathway is mediated by two multisubunit protein complexes of the chloroplast envelope. These were named the Toc (Translocon on the outer envelope of chloroplasts) and the Tic (Translocon on the inner envelope of chloroplasts) complex (Fig. 2). Translocation further requires a number of soluble factors in the cytosol, the intermembrane space, and the stroma. It also requires energy in form of ATP as well as GTP for regulation (Flügge and Hinz 1986; Pain and Blobel 1987; Schindler et al. 1987). At ATP concentrations of less than 50 µmol the precursor protein associates irreversible with the import machinery but is not fully translocated. This phase is referred to as the early import intermediate. At this stage the precursor protein has contact to several of the components of both Toc and Tic. More than 50 µmol ATP is necessary to complete translocation (Theg et al. 1989). During in vitro import experiments ATP can be either supplemented from the outside or regenerated by photosynthesis in the inside if chloroplasts are sufficiently illuminated during the experiment. Indeed for most precursor proteins no difference is observed, whether the in vitro import experiment is performed in the light or in the dark as long as enough ATP is present. It was thus assumed that ATP generation is the only function of light in the import process (Pain and Blobel 1987; Schindler et al. 1987; Theg et al. 1989; Pilon et al. 1992; Pilon et al. 1995). Hirohashi et al. (2001) demonstrated that light might as well have an regulatory function in import. They showed that two maize precursor proteins, pFdIII (precursor of plant-type ferredoxin isoprotein III) and pFNRII (precursor of ferredoxin-NADP$^+$ reductase isoprotein II), could be targeted into the chloroplast in the dark, when ATP was supplied. Yet they were detained in the intermembrane space when the import experiment was performed under illumination. This was confirmed for chloroplasts isolated from various sources, indicating that this phenomenon is not specific for a certain organism but might have been missed in the past (Hirohashi et al. 2001). By using chimeric proteins the authors established that the import behavior of pFdIII is determined by its transit peptide and not by the mature protein. While pFdI was shown to import equally well in the dark and the light, a chimer between the precursor of FdIII and the mature FdI was detained in the intermembrane space. Interestingly FdIII is considered a non-photosynthetic ferredoxin while FdI is involved in the transport of electrons through the electronic chain of the photosystems (Hase et al. 1991; Suzuki et al. 1991). Therefore import of chloroplast proteins might also be regulated based on their function.

So far most in vitro experiments support the concept of a single machinery that imports all kind of chloroplast proteins. Precursor proteins as different as pLHCP (precursor of light harvesting chlorophyll binding protein), pSSU, pFdI, pPC (precursor of plastocyanin), pAtpC (precursor of ATPase subunit C) or several of the subunits of the oxygen evolving system (pOEs) have been shown to compete with each other for import (Buvinger et al. 1989; Perry et al. 1991; Oblong and Lamppa 1992; Theg and Geske 1992). Unlike these in vitro data, recent in vivo experiments point toward a subsets of import components for recognizing distinct sets of chloroplast proteins (Jarvis et al. 1998; Bauer et al. 2000). Bauer et al. (2000) showed that *ppi2*, an *Arabid-*

opsis mutant lacking atToc160, is defective in importing proteins involved in the photosynthetic function of the chloroplast. Non-photosynthetic proteins such as the chorismate mutase, on the other hand, were still imported. The authors speculated that two recently discovered isoforms of atToc160, i.e. atToc120 and atToc132, can substitute for atToc160 in the recognition and hence translocation of some but not all chloroplast proteins.

4.1 Cytosolic Factors and the Guidance Complex

The import of many proteins into isolated chloroplasts can be achieved without the addition of any cytosolic factors. The import apparatus itself and its regulatory components seem to be predominantly integrated or tightly associated with the chloroplast envelope. First evidence for the involvement of soluble cytosolic proteins in chloroplast import arose from experiments with Lhcb1 (light harvesting chlorophyll binding protein 1). Heterologously expressed pLhcb1 could be imported into isolated chloroplasts only in the presence of soluble proteins obtained from leaf extract. Further experiments showed that Hsp70 (heat shock protein of 70 kDa), a cytosolic chaperone, was one of the factors that stimulated pLhcb1 import but it was not clear whether the requirement was owned to the specific nature of this highly hydrophobic protein. By now it is clear that chloroplast import generally employs cytosolic chaperones of the Hsp70 class. This feature is shared with import into mitochondria and the ER (Zimmermann et al. 1988; Waegemann et al. 1990). Over 70% of all chloroplast-targeted precursor proteins contain a putative binding site for DnaK, the *E. coli* Hsp70, within their transit sequence (IveyIII et al. 2000), indicating that interaction with a chaperone is feasible. This is further supported by experimental data that observed a direct interaction between precursor proteins and plant Hsp70 homologues (Rial et al. 2000). Chaperone interaction with precursor proteins occurs in a direct and supposedly unspecific manner (Ellis and van der Vies 1991). Its major purpose is most likely to prevent precursor protein aggregation in a highly bustling environment such asthe cytosol. They do not impart any specificity to the transfer.

A feature specific to chloroplast import is the phosphorylation of major chloroplast preproteins by a cytosolic kinase of unknown nature. Phosphorylation occurs on a specific serine or threonine residue located within the transit sequence (Waegeman and Soll 1996) and has been shown exclusively for chloroplast import. Precursor phosphorylation is not involved in import into mitochondria or other cell compartments. Interestingly, phosphorylated precursor proteins cannot be translocated into the chloroplast. The import is arrested at an early stage and can proceed only after dephosphorylation. A consensus motif for the phosphorylation site was deduced from several proteins including pSSU, pLhcb1, pOE23, and pOE33 (Waegemann and Soll 1996). The phosphorylation motif displayed sequence similarities to 14-3-3 phosphopeptide binding motifs which led to the idea of a possible interaction between 14-3-3 proteins and chloroplast precursor proteins (Fig. 1+2). 14-3-3 proteins are an ubiquitous eukaryotic protein family with regulatory and chaperone function. A cytosolic 14-3-3 protein is part of the multi-protein complex called mitochondrial import stimulation factor (Hachiya et al. 1995; Pfanner and Geissler 2001). May and Soll (2000) established a specific interaction between 14-3-3 proteins and the phosphorylated precursor proteins pSSU and pFdI. They identified a so-called guidance complex composed of 14-3-3, Hsp70 and possibly other proteins. Interaction of precursor protein with the guidance complex is strongly enhanced by transit peptide phosphorylation and greatly increases the import competence of the bound precursor protein (May and Soll 2000). The authors also showed that this interaction is specific for chloroplast import. When different precursor proteins were translated in either homologous or heterologous translation systems (wheat germ or reticulocyte lysate), mitochondrial precursor proteins did not interact with 14-3-3 proteins from wheat germ. Nor did chloroplast precurosr proteins interact with 14-3-3 proteins present in the reticulocyte lysate (May and Soll 2000). A precursor protein bound to the guidance complex engages the general import pathway after arrival at the outer envelope. Dissociation from the guidance complex is required prior to translocation and dephosphorylation of the precursor protein might play a role in the dissociation process.

4.2 The Toc Complex

To our present knowledge the Toc complex consists of at least 4 different proteins (Fig. 2). Most of the biochemical identification and characterization of these proteins has been accomplished in pea and the peaToc components, named according to their molecular weight (Schnell et al. 1997), include Toc159 (formerly Toc86), Toc75, Toc64, and Toc34. The completion of the *Arabidopsis* genome sequence has identified many homologues to these proteins and additional genetic approaches that were not possible in pea have added since to our knowledge about the function of these components.

4.2.1 Toc159

All Toc subunits are integral elements of the outer envelope membrane. Toc 159 has no classical membrane spanning domain but was shown to anchor into the outer envelope by its C-terminus. The largest portion of the protein, more than 100 kDa, is exposed to the cytosol (Hirsch et al. 1994; Kessler et al. 1994; Bölter et al. 1998). This cytosol exposed domain is highly susceptible to proteolytic cleavage and the protein easily breaks down to a 86 kDa fragment.

Further proteolytic degradation of intact chloroplasts leaves only the 52 kDa membrane anchor domain intact. Thus the full size of Toc159 was not discovered for many years. Toc159 can be cross-linked to Toc75, Toc34, and Toc64 during in vitro import experiments (Hirsch et al. 1994; Perry and Keegstra 1994; Kouranov and Schnell 1997; Sohrt and Soll 2000). It can also be cross-linked to precursor proteins especially under conditions that favor binding but not translocation (Perry and Keegstra 1994; Kouranov and Schnell 1997). This is one of the reasons why Toc159 has been implicated as a chloroplast import receptor. Also proteolytic degradation to Toc86 was shown to affect precursor binding rather than protein translocation (Chen et al. 2000).

Toc159 can be divided in three structural domains. The N-proximal part of the protein is highly acidic, containing a high percentage of aspartic or glutamic acid residues. It has been speculated that this domain can bind electrostatically to precursor proteins due to an overall positive net charge of the transit peptide (Bölter et al. 1998). A similar interaction between the transit sequence and an acidic receptor has been described for mitochondria (Komiya et al. 1998). In *Arabidopsis* three homologues of Toc159 are described, named atToc160, atToc132, and atToc120. All three of them resemble Toc159 in its C-terminal part but the two smaller proteins have len homology at the N-terminus. Nevertheless the strong acidity is preserved in the N-terminus of all three proteins (Bauer et al. 2000). The central part of Toc159 comprises highly conserved GTP-binding motifs and the functionality of this domain as a GTPase has been demonstrated in vitro (Seedorf et al. 1995; Kouranov and Schnell 1997; Chen et al. 2000, Sveshnikova et al. 2000a). GTP hydrolysis by Toc159 might be involved in the transfer of the precursor protein to other Toc subunits. Specific antibodies raised against Toc159 inhibit precursor binding to the Toc machinery and translocation into the chloroplasts indicating a direct function of Toc159 in precursor binding. Toc159 is furthermore phosphorylated by a 70 kDa outer envelope kinase (Fulgosi and Soll 2002). Toc159 appears to be the specific substrate of this kinase and phosphorylation might play a role in import regulation. A recent article by Hiltbrunner et al. (2001) also suggests that Toc159 exists in abundance in a soluble cytosolic form waiting to be targeted by Toc34 into the outer membrane of the chloroplasts. Further experimental evidence is required to corroborate this hypothesis and elucidate the function of such a soluble pool of Toc159.

4.2.2 Toc34

Toc34 shares a certain sequence homology to Toc159. They both contain a very similar GTPase domain that extents further than the conserved GTP-binding motifs (Kessler et al. 1994; Seedorf et al. 1995; Kouranov and Schnell 1997; Sveshnikova et al. 2000a). Toc34 is anchored to the outer envelope by a small membrane spanning domain close to its C-terminus while the major part of the protein protrudes into the cytosol. Like Toc159, it has been implicated as a receptor for precursor binding. Conditions that lead to translocation of precursor protein through the import machinery result in a loss of binding to Toc34. Thus the role of Toc34 in translocation seems to be early in the process and only transient. In vitro experiments by Sveshnikova et al. (2000a) using a truncated soluble Toc34 protein indicate that the interaction with precursor proteins is a complex process involving both GTP-binding and phosphorylation/dephosphorylation of Toc34. Phosphorylation of Toc34 is provided by a 98 kDa kinase present in the outer envelope membrane (Fulgosi and Soll 2002). Likewise this kinase seems to be specific for Toc34. The intricate regulation of Toc34 could be important for a differential control of protein import into chloroplasts. Further support for a direct involvement of Toc34 in chloroplast import comes from an analysis of the *Arabidopsis* mutant *ppi1*. Mutant plants lack functional atToc33, the more abundant of the two *Arabidopsis* Toc34 homologues, and have a reduced import efficiency (Jarvis et al. 1998).

Just anew the 2Å crystal structure of the soluble domain of Toc34 from pea has been described (Sun et al. 2002). It corroborated the similarity of Toc34 to other GTPases. The crystallized protein was present as a dimer, even though only 25% of heterologously expressed soluble peaToc34 is dimerized in a buffered solution. The Toc34 dimer contained both GDP and Mg-ions and GDP bound to both subunits was required for dimerization. The authors suggest a model by which one Toc34 subunit can act as a GTPase activation protein for the other subunit in a dimer and visa versa. These data present further evidence for a regulative function of Toc34 in chloroplast import.

4.2.3 Toc75

Toc75 is the most abundant protein of the chloroplast outer envelope. It interacts with precursor proteins and with Toc159 and Toc34. Especially Toc75 and Toc34 are in close physical proximity and can be covalently connected to each other by a reversible intermolecular disulfide bond. Interaction of Toc75 with Toc159 and Toc34 occurs even in the absence of precursor (Perry and Keegstra 1994; Schnell et al. 1994). It is generally believed that Toc75 is the translocation pore of the Toc complex (Perry and Keegstra 1994; Schnell et al. 1994; Tranel et al. 1995; Hinnah et al. 1997). Toc75 contains multiple membrane spanning regions predicted to form a β-barrel structure (Hinnah et al. 1997; Sveshnikova et al. 2000b). These would account for the ability of Toc75 to form a translocation pore. Hinnah et al. (1997) confirmed that recombinant Toc75 reconstituted in liposomes acts as a voltage-gated peptide-sensitive channel. Its pore size is at least 8–9 Å in diameter (Hinnah et al. 1997) but more likely the opening is as large as 20 to 25 Å (Hinnah et al. 2002 in press). This is in good accordance with a predicted function of Toc75 as a translocation pore and also supports the concept that precursor proteins have to be partially unfolded in order to translocate. Recombinant Toc75 that was reconstituted in liposomes reacted specifically with pSSU but not with mature SSU; an indication that the protein itself is capable of recognizing precursor proteins.

4.2.4
Toc64

The latest addition to the Toc-family is Toc64. The protein is part of isolated Toc complexes as evidenced by silver staining and Western blot analysis (Sohrt and Soll 2000). Toc64 is an integral membrane protein with a large cytosol-exposed C-terminal domain. The authors suggest a specific interaction between Toc64 and the guidance complex. Toc64 might be the primary docking site for precursor proteins that are directed to the chloroplast by the guidance complex. Notably Toc64 contains a domain commonly found in amidases. The domain is mutated at a position conserved in the reaction center of other amidases which could account for the lack of amidase activity. Toc64 further contains three tetratricopeptide repeats (TPR motifs). TPR motifs are found in many proteins including several subunits of the mitochondrial import apparatus and also in a peroxisomal import receptor. They are believed to facilitate protein-protein interaction and might be the interface between Toc64 and the guidance complex.

4.3
The Intermembrane Space

Precursor proteins translocate into the chloroplasts simultaneously through the outer and inner envelope (Alefsen et al. 1994; Schnell et al. 1994). Even though, the process is assisted by components of the intermembrane space. An Hsp70-type protein was found associated with the outer envelope membrane but facing the intermembrane space (Fig. 2). It can be co-immunoprecipitated together with pSSU during import and it was suggested that it interacts with the precursor proteins during their transit through the translocon (Marshall et al. 1990; Waegemann and Soll 1991; Schnell et al. 1994). The exact role of this chaperone in the import process is still elusive.

4.4
The Tic Complex

While the major components of the Toc complex and their function is reasonably well understood, much less is known about the exact composition of the Tic complex. The view on components of the Tic complex has been controversial and several, often divergent components have been identified by different groups as part of this translocation machinery. In several cases an agreement exists about the involvement of a protein, but its function is still debated. The reason for these sometimes controversial results might be explained by the presence of distinct sets of Tic complexes. All in all, a minimum of six different proteins, Tic110, Tic62, Tic55, Tic40, Tic22, and Tic20, seem to represent the bonafide translocon constituents (Fig. 2) (Ko et al. 1995; Kessler and Blobel 1996; Lübeck et al. 1996; Caliebe et al. 1997; Kouranov and Schnell 1997; Stahl et al. 1999).

4.4.1 Tic110

The largest component of the Tic complex is Tic110 (Kessler and Blobel 1996; Lübeck et al. 1996). It has been proposed that Tic110 interacts with stromal chaperones and attracts those to the import apparatus. It was shown to interact with both stromal Hsp100 and Cpn60 (see below) even though interaction with the latter may only take place in the presence of precursor protein (Kessler and Blobel 1996; Nielsen et al. 1997; Kouranov et al. 1998). At the same time an interaction of Tic110 with components of the Toc complex and the intermembrane space have been proposed. A recent nublication by Heins et al. (2002, inpress) shows that liposome reconstituted Tic110 can form an aqueous channel. Tic110 therefore represents at least part of the translocation pore of the Tic complex, a function previously unassigned for in this translocon. The pore is most likely formed by a β-barrel like structure, though Tic110 also contains at least one predicted transmembrane α-helix. Tic110 seems to expose large domains to both sides of the membrane which could be responsible for the proposed roles mentioned above in addition to its role as a protein translocation channel.

4.4.2 Tic62

When intact Tic complexes are purified from pea inner envelope fractions by Blue Native polyacrylamid gel electrophoresis, Tic62 is found associated with Tic110 and Tic55 (Küchler et al. submitted). The protein can be co-precipitated by antibodies raised against Tic110 and Tic55 further supporting the close proximity of these proteins. The deduced amino acid sequence of the *Arabidopsis* homologue, atTic62, contains a nicotinamide-dinucleotide-binding site close to the N-terminus followed by a putative hydrophobic membrane anchor. It was furthermore shown for Tic62 from pea that it can bind nicotinamide-dinucleotides (Küchler et al. submitted). The C-terminal part of Tic62 comprises several highly conserved, repetitive sequence moduls. Via these modules Tic62 was shown to associate with FNR at the inner envelop of the chloroplasts. In toto the features of Tic62 argue in favour of a role in redox regulation of chloroplast import. A specific influence of the redox state of the chloroplast on import had been suggested previously (Caliebe et al. 1997; Hirohashi et al. 2001).

4.4.3 Tic55

Tic55 is an integral membrane protein but its precise topology has not been established yet. The protein contains two membrane-spanning domains at its C-terminus and the major part is thought to extend into the stroma of the chloroplast. Tic55 can be found as part of a complex comprising precursor protein, Tic110, and components of the Toc complex. It contains a predicted Rieske-type iron-sulfur center and a mononuclear iron binding site (Caliebe et al. 1997). Iron-sulfur clusters might be used to sense changes in the redox state of the chloroplast and thus Tic55 might act as a regulator for import at the stage of the Tic complex. This hypothesis would agree with a possible import regulation in correlation to environmental conditions, e.g. light, and other developmental factors as discussed before.

4.4.4 Tic40

Tic40 (formerly also called Com44/Cim44 or Toc36) was originally characterized as a protein (or protein family) present in both the inner and outer envelope membrane (Ko et al. 1995). Stahl et al. (1999) showed that only a single Tic40 protein exists which is located entirely in the inner envelope and is a functional component of the Tic complex. The protein can be cross-linked to Tic110 and to precursor protein arrested during import. Tic40 contains a membrane-spanning domain at its N-terminus and has a predicted Hsp70 binding site in its C-terminal stroma-exposed part. The protein was thus implicated in the recruiting of stromal chaperones to the import machinery but this implication is not yet supported by experimental data.

4.4.5 Tic22 and Tic20

Tic20 and Tic22 are two further components of the Tic complex. They both interact with precursor protein during import, most likely in a sequential order. Tic 20 is a small integral protein of the inner envelope with three predicted transmembrane domains (Kouranov et al. 1998). It was shown to interact with a complex of Tic110 and Toc components but neither with Tic110 alone nor with Tic55 or Tic 22. Tic22 displays similar interaction features as Tic20 but is only peripherally associated with the inner envelope on the side facing the intermembrane space (Kouranov and Schnell 1997; Kouranvo et al.

1998). Tic22 is quite hydrophilic and was implicated as a receptor for precursor proteins coming from the Toc complex. Alternatively a function in the establishment of contact sites between inner and outer envelope is discussed (Kouranov et al. 1998).

4.5 Stromal Factors

Several soluble factors in the stroma assist the membrane-integrated Toc and Tic machinery in the processing of precursor protein after translocation (Fig. 2). The most important stromal protein involved in chloroplast import is the stromal processing peptidase (SPP). SPP cleaves the transit peptide of precursor proteins to release the mature protein (Oblong and Lamppa 1992; VanderVere et al. 1995; Richter and Lamppa 1998). SPP is a metallo endopeptidase containing the signature zinc-binding motif of this protein class. Isolated or heterologously expressed SPP is able to process precursor protein in vitro. This indicates that its function is independent from the import machinery as well as any other stromal factors. The mechanism by which SPP recognizes the transit peptide and the correct cleavage site is not very well understood. It has been shown that it cleaves the entire transit peptide in one step. The peptide is then degraded further, most likely by a second ATP-dependent protease that is not yet identified (Richter and Lamppa 1998). In vivo the degradation of the transit peptide is so fast that intermediate states are difficult to detect in experimental set-ups.

Stromal factors that interact with the imported protein furthermore include diverse members of the chaperone family, e.g. a 93 kDa chloroplast homologues of Hsp100 called ClpC or Hsp93 (Akita et al. 1997; Kouranov et al. 1998). Recent data suggest that Hsp93 is a constitutive part of the import machinery (Jackson-Constan et al. 2001). Its putative role is the binding of precursor proteins upon entering the chloroplast stroma. It thus prevents their backwards movement and might pull precursor proteins through the translocation pore, a function similar to Hsp70 in mitochondria and the ER (Herrmann and Neupert 2000; Strub et al. 2000). Other potential chaperones include homologues of Hsp60 (Cpn60) and Hsp70, which have been identified inside the chloroplasts (Marshall et al. 1990). Cpn60 was shown to interact with several precursor proteins directly upon their import into the stroma (Lubben et al. 1989; Tsugeki and Nishimura 1993). Additional to their direct involvement in translocation, chaperones participate in the correct folding of stromal proteins and prevent premature folding of proteins that are translocated further into other chloroplast sub-compartments.

4.6 Protein Import in *Arabidopsis*, a Genomic Approach

Most biochemical studies on chloroplast import have been performed on pea. Based on these in vitro studies many of the protein components of the chloroplast import machinery and their encoding genes have been identified. However, since the completion of the *Arabidopsis* genome sequencing project, this weed has become the genetic model organism for many areas of plant research (The *Arabidopsis* Genome Initiative 2000). Genomic approaches have been performed, in order to identify homologues for all subunits of the import machinery in *Arabidopsis* (Jarvis et al. 1998; Bauer et al. 2000; Jackson-Constan and Keegstra 2001). These studies detected homologues to all previously identified import subunits in at least a single copy in the *Arabidopsis* genome. Evidently the composition of the import machinery is conserved between pea and *Arabidopsis* and most likely in all higher plants. More often not only a single but several *Arabidopsis* gene homologues were found. In some cases the homology extents only to parts of a protein and thus might simply be an example for exon shuffling (Jackson-Constan and Keegstra 2001); more often these genes seem to encode true isoforms. Following-up biochemical and genetic analyses have determined for some of these genes whether all isoforms are expressed and functional (Jarvis et al. 1998; Bauer et al. 2000).

As mentioned above three genes with homologies to Toc159 could be identified, atToc160, atToc132, and atToc120. Northern blot analyses indicated that atToc160 is by far the most abundant of the three both in greening as well as in etiolated seedlings (Bauer et al. 2000). It is therefore considered the functional homologue of Toc159 and its disruption affects predominantly the translocation of proteins involved in the photosynthetic process. While the function of atToc132 and atToc120 is not yet clear, they might play a specific role in translocation of non-photosynthetic proteins. Toc75 also has three potential homologues in the *Arabidopsis* genome (Jackson-Constan and Keegstra 2001). No further analysis has been done on these proteins so that nothing is known about the function of the different isoforms. Only one of the three homologues is represented by an expressed sequence tag (EST) and thus it might be the only active form of Toc75. Two isoforms have been identified for Toc 34, i.e. atToc34 and atToc33 (Jarvis et al. 1998). Both of these isoforms are expressed in vivo in *Arabidopsis* (Jarvis et al. 1998; Gutensohn et al. 2000) but to different levels. atToc33 is the more abundant of the two and therefore likely the functional homologue of Toc34. *Arabidopsis* mutant plants of atToc33 and atToc34 import chloroplast proteins with reduced efficiency, even though their phenotype is much less severe than in mutants of atToc160. Multiple isoforms have been found for several other Toc and Tic components but in most cases the function of these isoforms is not clear. Other proteins, like Tic110, have only a single homologue in the *Arabidopsis* genome indicating that one polypeptide is sufficient for the import of all different type of proteins at all times (Jackson-Constan and Keegstra 2001).

The presence of so many multiple isoforms of import components has thrown some doubt on the theory that a single, ubiquitous import apparatus is responsible for the import of all type of precursor proteins in all different tissues. Further experimental evidence is needed to enlighten us on the reason for this redundancy of import components. More important, it remains to be shown whether this redundancy is a specific situation in *Arabidopsis* or whether it also extents to other plants. *Arabidopsis* offers the possibility for genetic approaches that are not open in a plant like pea for which many genetic tools are not established (Jackson-Constan and Keegstra 2001). This includes the generation and screening of 'knock-out' mutants, as well as the expression of sense and antisence genes. However, while the genetic analysis of the *Arabidopsis* genome has opened up new revenues for research, further experimental data have to elucidate these new questions.

5 Intraplastidal Sorting

For stroma located proteins the chloroplast quest is fulfilled once they are translocated through the Toc and Tic complex and processed by SPP. If required they are affixed with co-factors or assembled into larger protein complexes. Other proteins have to be routed further to their final destination in one of several chloroplast sub-compartments. These include proteins located in the envelope membrane but it centrally concerns all proteins of the thylakoid membrane and the thylakoid lumen. The thylakoid network is the dominating structure inside of mature chloroplasts. This interconnected membrane system builds a separate compartment within the chloroplasts. It is the place where the two photosystems and the oxygen-evolving system are located. Thylakoid membranes have evolved in the cyanobacterial ancestor of chloroplast and translocation into and across this membrane belongs to the bacterial heritage of chloroplasts.

5.1 Insertion into the Inner Envelope Membrane

All nuclear-encoded inner envelope proteins identified so far are synthesized with a transit peptide and engage the general import pathway (Flügge et al. 1989; Knight and Gray 1995; Li et al. 1992; Lübeck et al. 1997). The information for further translocation into the inner envelope membrane is not specified by a cleavable signal sequence. Instead this information seems to lie within the mature part of the protein (Li et al. 1992; Knight and Gray 1995). At least Tic110 was shown to completely translocate into the stroma (Lübeck et al. 1997). Only after the removal of the transit peptide is the protein relocated into the inner envelope via a thus far unidentified mechanism. Other inner envelope proteins have not been studied in detail but different pathways can be envisioned. They might only engage part of the import machinery, for instance they could translocate via the Toc complex into the intermembrane space and insert from there into the inner envelope. Alternatively they might as well enter the Tic complex and integrate laterally into the inner envelope membrane.

5.2 Thylakoid Targeting

Many of the chloroplast proteins are localized on the thylakoid membrane or inside the thylakoid lumen (Herrmann 1999). These include nuclear-encoded proteins as well as proteins encoded by the chloroplast genome. The former are translocated into the chloroplast by the general import pathway and then further routed to the thylakoids (Fig. 3). They can be divided into two groups; intrinsic proteins of the thylakoid membrane and proteins that are translocated into the thylakoid lumen. The latter are synthesized with a bipartite presequence (Dalbey and Robinson 1999). The N-proximal part of the presequence resembles the transit peptides of other chloroplast targeted proteins and is utilized to engage the general import pathway. The transit peptide is cleaved by SPP and the C-proximal part of the presequence guides the protein across the thylakoid membrane.

Thylakoid insertion and translocation is achieved by several pathways, all of which resemble various protein export and insertion pathways known from eubacteria and/or the cytosol of eukaryotic cells (Fig. 3). To date four different pathways have been identified: the signal recognition particle (SRP) pathway, the ΔpH-dependent or TAT-pathway, the Sec-pathway, and spontaneous insertion (for a more specialized review see: Dalbey and Robinson 1999; Eichacker and Henry 2001; Mori and Cline 2001; Schleiff and Klösgen 2001). While all of these pathways contain unique components and translocate different sets of proteins there is growing evidence that certain elements might be shared between them.

5.2.1 The Sec-Pathway

The Sec-pathway is an universal translocation pathway found in all domains of life; bacteria, archaea and eukaryotes. It was originally thought to specifically transport secretory proteins across the bacterial plasma membrane but has been shown in the meanwhile to also assist in the insertion of membrane proteins. In bacteria, e.g. *E. coli,* three integral membrane proteins, SecE, Y, and G, probably form a pore through which translocation occurs (Dalbey and

Robinson 1999). A fourth, peripheral subunit, SecA, is located at the cytoplasmic site of the plasma membrane. SecA is an ATPase and ATP hydrolysis is required to drive translocation across the plasma membrane. Homologues to several of the bacterial Sec proteins have been found in cyanobacteria as well as in chloroplasts, e.g. cpSecA, cpSecE, and cpSecY (Fig. 3).

In chloroplasts the pathway is part of the translocation machinery that transport proteins into the thylakoid lumen or integrates them into the thylakoid membrane. Proteins translocated by the Sec pathway include the 33 kDa subunit of the oxygen evolving system (OES33), plastocyanin and PS1-F. The functionality of cpSecY and cpSecA has been demonstrated by transposon tagged mutagenesis and also by in vitro import experiments (Voelker and Barkan 1995; Haward et al. 1997; Schuenemann et al. 1999). In bacteria SecY is part of the translocation pore and cpSecY probably has the same function. Mutants of cpSecY display severe phenotypes with a near complete loss of thylakoid membranes (Roy and Barkan 1998). Fröderberg et al. (2001) demonstrated recently that the chloroplast SecE homologue, cpSecE, can functionally replace the *E. coli* SecE protein. This is a strong indication for functional homology to the bacterial protein and further supports the idea of a bacterial-like Sec-pathway that translocates thylakoid proteins in chloroplasts.

5.2.2 The ΔpH-Dependent or TAT-Pathway

The ΔpH-dependent pathway was first described for protein translocation across the thylakoid membrane, where it is utilized by a limited subset of thylakoid lumen proteins. Translocation via this pathway functions independent of soluble stromal factors (other than chaperones) and does not require energy in form of nucleoside triphosphates. Even though, translocation of these proteins has only be achieved in intact chloroplasts but not into isolated thylakoids. Its prerequisite is an intact proton gradient across the thylakoid membrane as indicated by its name (Mould and Robinson 1991; Klösgen et al. 1992). In general, proteins that are translocated by this pathway possess a hydrophobic signal sequence and have been characterized by a twin arginine translocation (TAT) motif that is located in the N-terminal part of the proteins. Proteins shown to engage the Tat-pathway include PSI-N, OES23 and OES16/17 (Chaddock et al. 1995; Mori and Cline 1998). A homologous pathway was since identified in bacteria where it is used to translocate a distinct group of proteins, i.e. those binding complex redox factors, into the periplasmatic space (Dalbey and Robinson 1999). In general the TAT-pathway is used to translocate folded proteins, e.g. proteins with co-factor assembly prior to membrane passage.

The *E. coli* TAT-pathway employs at least four different proteins named TatA/E, B, and C, all of which are integral membrane proteins. Homologues of all of these proteins have been identified in chloroplasts (Fig. 3). In *E. coli*, TatC is the major constituent of the translocation pore; the function of cpTatC is not yet shown. An analysis of *Arabidopsis apg2* (cpTatC) mutants revealed that the protein is essential for proper chloroplast development (Motohashi et al. 2001). Mutant plants were chlorophyll and carotenoid deficient, resulting in an albino phenotype. Their chloroplasts lacked thylakoid membranes and thus also membrane bound and associated proteins, including D1 or Lhcb1. It was assumed that these proteins are degraded since they cannot be assembled into the membrane. The composition of soluble stroma proteins, on the other hand, was not affected by the mutation (Motohashi et al. 2001). Chloroplast homologues of TAT subunits have been identified by genetic approaches. Mutant analysis of maize and *Arabidopsis* identified the *hcf106* (TatB) and the *tha4* (TatA/E) gene (Martienssen et al. 1989; Voelker and Barkan 1995). Disruption of any of these gene loci resulted in a severe loss of chloroplast structure and function. Recent studies by Cline and Mori (2001) suggested that cpTatC and the Hcf106 gene product both function as a precursor receptor. The precursor/cpTatC/cpTatB complex subsequently binds to Tha4 (TatA/E) and forms a translocation channel by structural rearrangement.

Only recently Molik et al. (2001) showed that the Rieske Fe/S protein, which is part of the cytochrom b_6/f complex, utilizes the TAT-pathway. It is the first integral membrane protein to use this pathway and also the first that does not contain a cleavable signal peptide. The protein furthermore has no twin-arginine motif and its insertion could be competed for by proteins that engage the Sec pathway. This is further evidence that at least these two pathways are not independent but either share certain components or communicate with each other.

5.2.3 The SRP Pathway

While the two pathways described above employ a presequence for thylakoid targeting there are also thylakoid proteins without any recognizable presequence. The information for thylakoid targeting must thus lie somewhere within the mature protein. Several of those proteins engage the signal recognition particle (SRP) pathway. SRP is a common pathway for membrane protein insertion. It is known from bacteria where it promotes insertion into the plasma membrane (Dalbey and Robinson 1999). It also occurs in the cytosol of eukaryotic cells where it is involved in insertion into and translocation across the ER membrane. The pathway is in principle conserved in different organisms but alterations in its composition do occur. Both, bacterial and cytosolic SRP pathways contain at least one protein, SRP54, and a ribonucleic acid. The chloroplast SRP pathway, involved in post-translational insertion of proteins contains at least two proteins, cpSRP54 and cpSRP43 (Fig. 3). So far no RNA moiety has been found associated with the SRP complex in chloroplasts. cpSRP54 has a high sequence homology to SRP54 proteins found in both bacteria and eukaryotes. cpSRP43 does not resemble any other known protein but it is postulated to function in determing the targeting activity of cpSRP54 (Groves et al. 2001).

All nuclear-encoded proteins inserted into the thylakoid membrane by cpSRP are part of the light harvesting chlorophyll protein family (Lhcp). The pathway is best studied for the insertion of Lhcb1 (Franklin and Hoffmann 1993; Li et al. 1995) but it is assumed that it is utilized by many other Lhcp's. Lhcb1 protein binds to the cpSRP complex directly after import through the general import pathway under formation of a so-called transit complex. In this conformation the otherwise hydrophobic protein is soluble in the aqueous stroma (Payan and Cline 1991; Li et al. 1995). Insertion of Lhcb1 into the thylakoid membrane requires GTP (Hoffmann and Franklin 1994), the plastidal cpSRP receptor FtsY (Kogata et al. 1999; Tu et al. 1999), and the integral membrane protein Alb3 (Moore et al. 2000) (Fig. 3). Alb3 acts as the integrase of this pathway and is the chloroplast homologue of the mitochondrial Oxa1p and the bacterial YidC protein. Alb3 is required for the membrane insertion of at least two further Lhcps (Lhcb4.1 and Lhcb5) but in contrast to bacteria it is also involved in the insertion of SRP-independent proteins (Woolhead et al. 2001). Experimental studies on Lhcb1 indicated at least two internal domains required for insertion (High et al. 1997; deLille et al. 2000). *Arabidopsis* mutants with a reduced content of cpSRP54 display a phenotype of yellow primary leaves with immature chloroplasts and a reduced chlorophyll content (Pilgrim et al. 1998). The mutation moreover affected proteins not targeted by the cpSRP pathway. Older leaves appeared normal despite a still reduced level of cpSRP54 suggesting that the lack of this protein can be bypassed (Pilgrim et al. 1998). This is in contrast to mutations in the ΔpH and the Sec-pathway which were shown to be lethal in maize (Voelker and Barkan 1995). Recent studies in the diatom *Odontella sinensis* and *Coscinodiscus granii* indicated that the SRP pathway is also present in certain complex plastids (Lang and Kroth 2001).

5.2.4 Spontaneous Insertion

Several chloroplast proteins can insert into isolated thylakoid membranes in the complete absence of soluble components of the stroma and even after preteatment of the membranes with proteases (Fig. 3). The same phenomenon had been observed for a subgroup of bacterial membrane proteins and due to the lack of any auxiliary components it was termed spontaneous insertion. CF_0II, PsaK, PsbX, and PsbW are proteins that have been shown to insert spontaneously into the thylakoid membrane. They all contain a single-membrane spanning domain and a cleavable signal sequence (Kim et al. 1999;

Mant et al. 2001; Woolhead et al. 2001). Other studies suggest spontaneous insertion for proteins with multi-membrane domains or of proteins without a cleavable signal sequence (Kim et al. 1999). Future investigations will have to clarify whether this insertion is indeed spontaneous or whether it occurs by a yet not identified protein-mediated pathway.

5.3 Intraplastidal Vesicle Transport

After translocation through the import machinery proteins have to travel to their final location inside the chloroplast. It has been shown for Lhcb1 that it binds to the cpSRP complex right after release from the import machinery (Payan and Cline 1991; Li et al. 1995). Supposingly the binding allows this highly hydrophobic protein to stay soluble in the aqueous stroma. For other hydrophobic proteins located in the thylakoid membrane their means of intraplastidal transport is not clear.

Moreover, several other elements of the thylakoid membrane, including membrane lipids, are synthesized outside of the organelle or at the chloroplast envelope. All of these components have to be transported through the stroma since in mature chloroplasts no connection exists between the envelope membrane and the thylakoid system. Vesicle-like structures appear in the stroma of higher plant chloroplasts under certain conditions (Morre et al. 1991; Kroll et al. 2001; Westphal et al. 2001a,b). They have been observed since early electron microscopic studies and thought to be involved in membrane traffic. Westphal et al. (2001b) showed that the formation and fusion of these vesicles is affected by low molecular weight components in a fashion similar to cytosolic vesicle traffic. Vesicle transport is a common phenomenon in eukaryotic cells, yet it has not been shown in any prokaryotic organisms, including cyanobacteria. This intraplastidal vesicle transport system might well be partially of eukaryotic descent and was set up inside the chloroplast subsequent to the development of the organelle. Such a vesicle transport system would explain several problems concerning thylakoid formation and maintenance. Nevertheless its composition and the true cargo transported in these vesicles remains to be revealed.

6 Protein Translocation into Complex Plastids

Complex plastids still contain a circular genome which encodes a limited set of proteins involved in transcription and translation and some of the components involved in their photosynthetic function. The apicoplast genome appears to encode even less proteins, most of which are required for transcription and translation. Much less is known about the total protein content of apicoplasts compared to chloroplasts but at least around 800 proteins are estimated to be translocated from the cytosol into the organelle (Waller et al. 2000). The number of proteins imported into photosynthetically active complex plastids is probably similar to chloroplasts.

Since complex plastids are surrounded by three or even four membranes, the import of nuclear encoded plastidal proteins is more complex than import into ordinary chloroplasts. Morphological studies revealed that the outer membrane of complex plastids from cryptomonads, patophytes and heterokons bears ribosomes and is continuos with the ER. Thus Gibbs (1978) suggested that nuclear-encoded proteins are translocated co-translationally through this membrane (called CER, chloroplast plastid ER), a theory that has been supported by further evidence since first proposed. Indeed it appears by now that the endomembrane system is the entrance way into all complex plastids. Diatoms, cryptomonads, apicomplexa and Euglena do not contain a continuum between the ER and the outer plastid membrane. Yet it is believed that in these organisms precursor proteins travel through the ER to the plastid by means of secretory vesicle traffic.

6.1 Bipartite Presequences

Targeting into all complex plastids is achieved by a bipartite N-terminal presequence composed of two domains. The N-proximal part is a signal peptide. It consists of a short hydrophobic domain followed by a specific cleavage motif. The homology of this peptide to signal peptides of ER targeted proteins was the first strong evidence that a pathway via the ER is involved in translocation into complex plastids. The signal peptide is followed by a transit peptide similar in its characteristics to known chloroplast targeting sequences (see chapter 3.1.1). In vivo experiments using GFP-fusion proteins as well as in vitro import assays demonstrated that the bipartite sequence is indeed essential for targeting to and import into the apicoplast (Waller et al. 2000). It was furthermore shown that the transit sequence of plastid targeted proteins from *Euglena*, apicomplexa, cryptomonads and heterokonts are able to facilitate import of proteins into ordinary chloroplasts of higher plants. This indicates a close phylogenetic relationship to transit peptides of algae and higher plants and supports the notion that the chloroplast import machinery was already developed in the phototrophic eukaryote prior to secondary endosymbiosis.

6.2 Targeting via ER and the secretory pathway

Translocation over the two innermost membranes (the envelope of the original chloroplast) is presumed to follow the same general import pathway as described for ordinary chloroplasts. More interesting is the question how translocation across the one or two outermost membranes is achieved.

It is believed that for all complex plastids routing to and translocation over the outermost membrane takes places through the ER and the secretory pathway. In CER containing organisms proteins will be co-translationally transferred into the periplastidal space. The signal peptide is cleaved and the protein is further routed to the chloroplast. In organisms without CER the signal peptide allows entrance of the organelle-targeted protein into the ER and further into the secretory pathway. These proteins are transported by secretory vesicle traffic from the ER to the plastid. Fusion of the secretory vesicle with the outmost membrane of the complex plastids releases the protein into the periplastidal space. How exactly the routing to the organelle is accomplished is not known but it was shown for apicomplexa that recombinant proteins containing the signal peptide but lacking the transit sequence are indeed secreted out of the cell (Waller et al. 2000). Several studies on algae containing complex plastids have supported this transport pathway (Bhaya and Grossman 1991; Sulli and Schwartzbach 1995; Lang et al. 1998).

A complete enigma remains the translocation across the second outermost membrane in complex plastids surrounded by four membranes. Potentially this membrane contains pores or even a translocation machinery that allows proteins to cross over. Alternatively, a vesicle transport from this membrane to the outer envelope membrane is discussed. Since these vesicles would fuse with the outer envelope of the plastid they would release their cargo into the intermembrane space and translocation into the plastid would require only the Tic complex. So far their is no experimental evidence supporting any of these pathways.

7
Evolution of the Import Apparatus

The necessity for a sophisticated machinery that promotes and regulates import into chloroplasts arose after the endosymbiotic event that created this organelle in concert with the translocation of genes into the nucleus of the host cell. Even though many components of the import machinery have been identified, their evolutionary origin remains widely elusive. At the time of the endosymbiotic event that lead to the formation of chloroplasts the organisms involved had successfully tackled the problem of translocating proteins into or across membranes multiple times. Analogous translocation machineries were present in the cyanobacterial endosymbiot, e.g. thylakoid insertion and protein export, as well as in the mitochondriate host cell, e.g. protein translocation into and across the plasma membrane and mitochondrial import.

Several of the Toc and Tic components contain protein domains that are also found in other proteins (Table 2). This includes the GTPase domain of

Table 1. Phylogenetic origin of chloroplast import components in cyanobacteria. Putative homologues in *Synechocystis* and *Anabaena* were identified by Blast searches against the complete genomes at Kazuka Research Institute.

subunit	homologues in cyanobacteria gene number	e-value
Toc159	none	
Toc75	alr2269	$9e^{-44}$
	alr4893	$1e^{-34}$
	alr0075	$5e^{-25}$
	slr1227	$5e^{-36}$
Toc34	all4820	$8e^{-05}$
	slr1974	$6e^{-04}$
Toc33	none	
Toc64	none	
Tic110	none	
Tic62	alr2751	$1e^{-33}$
	alr2903	$7e^{-16}$
	sll1218	$3e^{-25}$
Tic55	alr5007	$2e^{-63}$
	all2097	$2e^{-39}$
	alr4354	$7e^{-39}$
	all7348	$2e^{-29}$
	slr1747	$2e^{-35}$
	sll1869	$4e^{-25}$
	sll1297	$9e^{-17}$
Tic40	none	
Tic22	slr0924	$6e^{-05}$
Tic20	all4804	$1e^{-10}$
	sll1737	$4e^{-09}$

Toc159 and Toc34 as well as the TPR motifs of Toc64. These homologies are restricted to those domains and do not extent further along the rest of the proteins. For two Toc components, Toc75 and Tic55, true cyanobacterial homologues can be found (Table 1). The e-values indicate an evolutionary relationship, even though the function of cyanobacterial synToc75 and synToc55 is not clear. Bölter et al (1998) showed that synToc75 resides in the outer cyanobacterial membrane and heterologously expressed protein forms a high conductance channel with affinity for polyamines and peptides. This finding makes it possible that Toc75 was recruited into the import machinery from a preexisting channel-forming protein. The evolution of a fixed mem-

Tabelle 2. Phylogenetic relationship of import components to other proteins. Homologous proteins and protein domains were identified by PSI-Blast searches agains the non-redundant GenBank at NCBI

subunit	homology to proteins/domains
Toc159	Toc34 GTP binding domain
Toc75	none
Toc34	Toc33 Toc159 GTP binding domain
Toc33	Toc34
Toc64	amidases, aminotransferases TPR motif
Tic110	none
Tic62	ycf39 short chain dehydrogenases
Tic55	lethal leaf spot 1 cell death suppressor protein oxygenases Rieske-type iron-sulfur center
Tic40	Hsp binding domain
Tic22	none
Tic20	none

brane topology for synToc75/Toc75 prior to the endosymbiotic event would also explain the unusual insertion pathway of this outer envelope protein (see 3.2.1). Tic55 shares homologies to several cyanobacterial proteins. These include a notable similarity to a 'cell death suppressor protein' that is also found in higher plants in form of the lethal leaf spot protein 1. Sequence homologies extent further to a number of oxygenases. This could be due to a conserved Rieske-iron center present in Tic55. The significance of this similarity is not known. Tic62 has N-terminally significant sequence homology to putative proteins in both *Synechocystis* and *Anabaena*. The *Synechocystis* gene, *ycf39*, is also found in other organisms and contains a nicotinamide-dinuloetide-binding site. Toc62 is furthermore similar to a number of small chain dehydrogenases. As mentioned before the N-terminal part of Toc64 is highly homologuous to amidases. No amidase activity was detected by Sohrt and Soll (2000) and it is not known whether this domain plays a role in the function of Toc64. Alltogether it seems that very little of either the cyanobacterial or the hosts protein equipment was used in order to create the chloroplast import apparatus. Instead most of the chloroplast import machinery was developed newly around a few preexisting elements when the need a rose.

8 Outlook

In the last ten years we and other groups have identified single subunits of the chloroplast protein translocon and started to elucidate their function. With the crystallization of Toc34 the first molecular structure of a translocon component has just been described. The next 10 years will probably see the solution of the structure of further subunits and maybe the whole translocon machinery. The molecular mechanism behind the action of translocation components will become clear. We will further start to understand the regulatory networks which direct single events of the process and how they connect and coordinate organellar function with cellular requirements.

Acknowledgements Work of the authors' laboratory was supported in part by the Deutsche Forschungsgemeinschaft, the Sonderforschungsbereich TR1 and Fonds der Chemischen Industrie.

Abbreviations

cpSEC	chloroplastic SEC pathway or component
cpSRP	chloroplastic SRP pathway or component
cpTAT	chloroplastic TAT pathway
DGDG	digalactosyldiacylglyceride
ER	endoplasmic reticulum
Hsp	heat shock protein
IE	inner envelope
Lhcp	light harvesting chlorophyll binding protein
MGDG	monogalactosyldiacylglyceride
OE	outer envelope
OEP	outer envelope protein
OES	oxygen evolving system
pAtpC	precursor of ATPase subunit C
pFdI	precursor of ferredoxin isoprotein I
pFdIII	precursor of plant-type ferredoxin isoprotein III

pFNRII	precursor of ferredoxin-NADP$^+$ reductase isoprotein II
PG	phosphatidylglycerol
pLhcb1	precursor of light harvesting chlorophyll binding protein 1
pOE	precursor of a subunit of the oxygen evolving system
pPC	precursor of plastocyanin
pSSU	precursor of SSU
SPP	stromal processing peptidase
SRP	signal recognition particle
SSU	small subunit of ribulose-bisphosphate carboxylase/oxygenase
TAT	twin arginine
Tic	translocon on the inner envelope of chloroplasts
Toc	translocon on the outer envelope of chloroplasts

References

Akita M, Nielsen E, Keegstra K (1997) Identification of protein transport complexes in the chloroplastic envelope membranes via chemical cross-linking. J Cell Biol 136:983–994

Alefsen H, Waegemann K, Keegstra K (1997) Analysis of the chloroplast protein import machinery. J Plant Physiol 144:339–345

Arabidopsis Genome Initiative (2000) Analysis of the genome sequence of the flowering plant *Arabidopsis thaliana*. Nature 408:796–815

Bartling D, Clausmeyer S, Oelmüller R, Herrmann RG (1990) Towards epitope models for chloroplast transit sequences. Bot Mag 2:119–144

Bauer J, Chen K, Hiltbunner A, Wehrli E, Eugster M, Schnell D, Kessler F (2000) The major protein import receptor of plastids is essential for chloroplast biogenesis. Nature 403:203–207

Bendich AJ (1987) Why do chloroplasts and mitochondria contain so many copies of their genomes? BioEssays 6:279–282

Bhaya D, Grossman A (1991) Targeting proteins to diatom plastids involves transport through an endoplasmic reticulum. Mol Gen Genet 229:400–404

Bölter B, May T, Soll J (1998) A protein import receptor in pea chloroplasts, Toc86, is only a proteolytic fragment of a larger polypeptide. FEBS Lett 441:59–62

Bruce B (1998) The role of lipids in plastid protein transport. Plant Mol Biol 38:223–246

Buvinger WE, Michel H, Bennett J (1989) A truncated analog of a pre-light-harvesting chlorophyll *a/b* protein II transit peptide inhibits protein import into chloroplasts. J Biol Chem 264:1195–1202

Caliebe A, Grimm R, Kaiser G, Lübeck J, Soll J, Heins, L (1997) The chloroplastic protein import machinery contains a Rieske-type iron-sulfur cluster and a mononuclear iron-binding protein. EMBO J 16:7342–7350

Cavalier-Smith T (1987) The simultaneous symbiotic origin of mitochondria, chloroplasts, and microbodies. Ann N Y Acad Sci 503:55–71

Cavalier-Smith T (2000) Membrane heredity and early chloroplast evolution. Trends Plant Sci 5:174–182

Chaddock AM, Mant A, Karnauchov I, Brink S, Herrmann RG, Klösgen RB, Robinson C (1995) A new type of signal peptide: central role of a twin-arginine motif in transfer signals for the delta pH-dependent thylakoidal protein translocase. EMBO J 14:2715–2722

Chen K, Li HM (1998) A mutant deficient in the plastid lipid DGD is defective in protein import into chloroplasts. Plant J 16:33–39

Chen K, Chen X, Schnell DJ (2000) Initial binding of preproteins involving the Toc160 receptor can be bypassed during protein import into chloroplasts. Plant Physiol 122:813–822

Cline K, Mori H (2001) Thylakoid DeltapH-dependent precursor proteins bind to a cpTatC-Hcf106 complex before Tha4-dependent transport. J Cell Biol 154:719–729

Dalbey RE, Robinson C (1999) Protein translocation into and across the bacterial plasma membrane and the plant thylakoid membrane. Trends Biochem Sci 24:17–22

deCastro Silva Filho M, Wieerst MC, Flügge UI, Chaumont F, Boutry M (1996) Different *in vitro* and *in vivo* targeting properties of the transit peptide of a chloroplast envelope inner membrane Protein. J Biol Chem 272:15264–15269

DeLille J, Peterson EC, Johnson T, Moore M, Kight A, Henry R (2000) A novel precursor recognition element facilitates posttranslational binding to the signal recognition particle in chloroplasts. Proc Natl Acad Sci USA 97:1926–1931

Dobberstein B, Blobel G, Chua NH (1977) *In vitro* synthesis and processing of a putative precursor for the small subunit of ribulose-1,5-bisphosphate carboxylase in *Chlamydomonas reinhardtii*. Proc Natl Acad Sci USA 74:1028–1085

Eichacker LA, Henry R (2001) Function of a chloroplast SRP in thylakoid protein export, Biochem Biophys Acta 1541:120–134

Ellis RJ, van der Vies SM (1991) Molecular chaperones. Annu Rev Biochem 60:321–347

Fichera ME, Roos DS (1997) A plastid organelle as a drug target in apicomplexan parasites. Nature 390:407–409

Fischer K, Weber A, Arbinger B, Brink S, Eckerskorn C, Flügge U-I (1994) The 24 kDa outer envelope membrane protein from spinach chloroplasts: Molecular cloning, in vivo expression and import pathway of a protein with unusual properties. Plant Mol Biol 25, 167–177

Flügge UI, Hinz G (1986) Energy dependence of protein translocation into chloroplasts. Eur J Biochem 160:563–567

Flügge UI, Fischer K, Gross A, Sebald W, Lottspeich F, Eckerskorn C (1989) The triose phosphate-3-phosphoglycerate-phosphate translocator from spinach chloroplasts: nucleotide sequence of a full-length cDNA clone and import of the in vitro synthesized precursor protein into chloroplasts. EMBO J 1:39–46.

Franklin AE, Hoffman NE (1993) Characterization of a chloroplast homolog of the 54-kda subunit of the signal recognition particle. J Biol Chem 268:22175–22180

Fröderberg L, Röhl T, van Wijk K-J, de Gier J-WL (2001) Complementation of bacterial SecE by a chloroplastic homologue. FEBS Lett 498:52–56

Froehlich JE, Benning C, Dormann P (2001) The digalactosyldiacylglycerol (DGDG) synthase DGD1 is inserted into the outer envelope membrane of chloroplasts in a manner independent of the general import pathway and does not depend on direct interaction with monogalactosyldiacylglycerol synthase for DGDG biosynthesis. J Biol Chem 276:31806–31812

Fulgosi H, Soll J (2002) The chloroplast protein import receptors Toc34 and Toc159 are phosphorylated by distinct protein kinases. J Biol Chem in press

Gibbs SP (1978) The chloroplast of *Euglena* may have evolved from symbiotic green algae. Can J Bot 56 2883–2889
Goldschmidt-Clermont M (1998) Coordination of nuclear and chloroplast gene expression in plant cells. Int Rev Cytol 177:115–180
Groves MR, Mant A, Kuhn A, Koch J, Dubel S, Robinson C, Sinning I (2001) Functional characterization of recombinant chloroplast signal recognition particle. J Biol Chem 276:27778–277786
Gutensohn M, Schulz B, Nicolay P, Flügge U-I (2000) Functional analysis of the two *Arabidopsis* homologues of Toc34, a component of the chloroplast protein import apparatus. Plant J 23:771–783
Hachiya N, Mihara K, Suda K, Horst M, Schatz G, Lithgow T (1995) Reconstitution of the initial steps of mitochondrial protein import. Nature 376:705–709
Hansmann P, Eschbach S (1990) Isolation and preliminary characterization of the nucleus and the nucleomorph of a cryptomonad, Pyrenomonas salina. Eur J Cell Biol 52:373–378
Haward SR, Napier JA, Gray JC (1997) Chloroplast SecA functions as a membrane-associated component of the Sec-like protein translocase of pea chloroplasts. Eur J Biochem 248:724–30
Hase T, Kimata Y, Yonekura, K, Matsumura T, Sakakibara H (1991) Molecular cloning and differential expression of the maize ferredoxin gene family. Plant Physiol 96:77–83
Herrmann RG (1999) Biogenesis and evolution of photosynthetic (thylakoid) membranes. Biosci Rep 19:355–365
Herrmann JM, Neupert W (2000) What fuels polypeptide translocation? An energetical view on mitochondrial protein sorting. Biochim Biophys Acta 1459:331-338
High S, Henry R, Mould RM, Valent Q, Meacock S, Cline K, Gray JC, Luirink J (1997) Chloroplast SRP54 interacts with a specific subset of thylakoid precursor proteins. J Biol Chem 272:11622–11628
Hiltbrunner A, Bauer J, Alvarez-Huerta M, Kessler F (2001) Protein translocon at the *Arabidopsis* outer chloroplast membrane. Biochem Cell Biol 79:629–635
Hinnah SC, Hill K, Wagner R, Schlicher T, Soll J (1997) Reconstitution of a chloroplast protein import channel. EMBO J 16:7351–7360
Hirohashi T, Hase T, Nakai M (2001) Maize non-photosynthetic ferredoxin precursor is mis-sorted to the intermembrane space of chloroplasts in the presence of light. Plant Physiol 125:2154–2163
Hirsch S, Muckel E, Heemeyer F, Heijne Gv, Soll J (1994) A receptor component of the chloroplast protein translocation machinery. Science 266:1989–1992
Hoffman NE, Franklin AE (1994) Evidence for a stromal GTP requirement for the integration of a chlorophyll a/b-binding polypeptide into thylakoid membranes. Plant Physiol 105:295–304
Ivey RA 3rd, Subramanian C, Bruce B (2000) Identification of a Hsp70 recognition domain within the rubisco small subunit transit peptide. Plant Physiol 122:1289–1299
Jackson DT, Froehlich JE, Keegstra K (1998) The hydrophobic domain of Tic110, an inner envelope membrane component of the chloroplastic protein translocation apparatus, faces the stromal compartment. J Biol Chem 273:16583–16588
Jackson-Constan DT, Keegstra K (2001) *Arabidopsis* genes encoding components of the chloroplastic protein import apparatus. Plant Physiol 125:1567–1576
Jarvis P, Chen L-M, Li H-m, Peto CA, Chory J (1998) An *Arabidopsis* mutant defective in the plastid general import apparatus. Science 282:100–103

Jarvis P (2001) Intracellular signalling: the chloroplast talks! Curr Biol 11:307–310
Kerber B, Soll J (1992) Transfer of a chloroplast-bound precursor protein into the translocation apparatus is impaired after phospholipase C treatment. FEBS Lett 306:71–74
Kessler F, Blobel G, Patel HA, Schnell DJ (1994) Identification of two GTP-binding proteins in the chloroplast protein import machinery. Science 266:1035–1039
Kessler F, Blobel G (1996) Interaction of the protein import and folding machineries in the chloroplast. Proc Natl Acad Sci USA 93:7684–7689
Klösgen RB, Brock IW, Herrmann RG, Robinson C (1992) Proton gradient-driven import of the 16 kDa oxygen-evolving complex protein as the full precursor protein by isolated thylakoids. Plant Mol Biol 18:1031–1034
Knight JS, Gray JC (1995) The N-terminal hydrophobic region of the mature phosphate translocator is sufficient for targeting to the chloroplast inner envelope membrane. Plant Cell 9:1421–32
Ko K, Ko ZW (1992) Carboxyl-terminal sequences can influence the in vitro import and intraorganellar targeting of chloroplast protein precursors. J Biol Chem 267:13910–13916
Ko K, Budd D, Wu C, Seibert F, Kourtz L, Ko ZW (1995) Isolation and characterization of a cDNA clone encoding a member of the Com44/Cim44 envelope components of the chloroplast protein import apparatus. J Biol Chem 270:28601–28608
Köhler S, Delwiche CF, Denny PW, Tilney LG, Webster P, Wilson RJ, Palmer JD, Roos DS (1997) A plastid of probable green algal origin in Apicomplexan parasites. Science 275:1485–1489
Kogata N, Nishio K, Hirohashi T, Kikuchi S, Nakai M (1999) Involvement of a chloroplast homologue of the signal recognition particle receptor protein, FtsY, in protein targeting to thylakoids. FEBS Lett 447:329–333
Komiya T, Rospert S, Koehler C, Looser R, Schatz G, Mihara K (1998) Interaction of mitochondrial targeting signals with acidic receptor domains along the protein import pathway: evidence for the 'acid chain' hypothesis. EMBO J 17:3886–3898
Kouranov A, Schnell DJ (1997) Analysis of the interactions of preproteins with the import machinery over the course of protein import into chloroplasts. J Cell Biol 139:1677–1685
Kouranov A, Chen X, Fuks B, Schnell DJ (1998) Tic20 and Tic22 are new components of the protein import apparatus at the chloroplast inner envelope membrane. J Cell Biol 143:991–1002
Kroll D, Meierhoff K, Bechtold N, Kinoshita M, Westphal S, Vothknecht UC, Soll J, Westhoff P (2001) VIPP1, a nuclear gene of Arabidopsis thaliana essential for thylakoid membrane formation. Proc Natl Acad Sci USA 98:4238–4242
Lang M, Apt KE, Kroth PG (1998) Protein transport into „complex" diatom plastids utilizes two different targeting signals. J Biol Chem 273:30973–30978
Lang M, Kroth PG (2001) Diatom fucoxanthin chlorophyll a/c-binding protein (FCP) and land plant light-harvesting proteins use a similar pathway for thylakoid membrane insertion. J Biol Chem 276:7985–7991
Li HM, Moore T, Keegstra K (1991) Targeting of proteins to the outer envelope membrane uses a different pathway than transport into chloroplasts. Plant Cell 3:709–717
Li HM, Sullivan TD, Keegstra K (1992) Information for targeting to the chloroplastic inner envelope membrane is contained in the mature region of the maize Bt1-encoded protein. J Biol Chem 267:18999–9004
Li HM, and Chen L-J (1996) Protein targeting and integration signal for the chloroplastic outer envelope membrane. Plant Cell 8:2117–2126

Li XX, Henry R, Yuan JG, Cline K, Hoffman NE (1995) A chloroplast homolog of the signal recognition particle subunit SRP54 is involved in the posttranslational integration of a protein into thylakoid membranes. Proc Natl Acad Sci USA 92:3789–3793

Lübeck J, Soll J, Akita M, Nielsen E, Keegstra K (1996) Topoplogy of IEP110, a component of the chloroplastic protein import machinery present in the inner envelope membrane. EMBO J 15:4230–4238

Lübeck J, Heins L, Soll J (1997) A nuclear-coded chloroplastic inner envelope membrane protein uses a soluble sorting intermediate upon import into the organelle. J Cell Biol 137:1279–86

Lubben TH, Donaldson GK, Viitanen PV, Gatenby AA (1989) Several proteins imported into chloroplasts form stable complexes with the GroEL-related chloroplast molecular chaperone. Plant Cell 1:1223–1230

Mant A, Woolhead CA, Moore M, Henry R, Robinson C (2001) Insertion of PsaK into the thylakoid membrane in a „Horseshoe" conformation occurs in the absence of signal recognition particle, nucleoside triphosphates, or functional albino3. J Biol Chem 276:36200–36206

Margulis L (1970) Origin of Eukaryotic Cells. New Haven, CT, USA; Yale University Press

Martin W, Herrmann RG (1998) Gene transfer from organelles to the nucleus: how much, what happens, and why? Plant Physiol 118:9–17

Marshall JS, DeRocher AE, Keegstra K, Vierling E (1990) Identification of heat shock protein hsp70 homologues in chloroplasts. Proc Natl Acad Sci USA 87:374–378

Martienssen RA, Barkan A, Freeling M, Taylor WC (1989) Molecular cloning of a maize gene involved in photosynthetic membrane organization that is regulated by Robertson's Mutator. EMBO J 8:1633–1639

May T, Soll J (2000) 14-3-3 proteins form a guidance complex with chloroplast precursor proteins in plants. Plant Cell 12:53–64

McFadden GI, Gilson PR, Douglas SE (1994) The photosynthetic endosymbiont in cryptomonad cells produces both chloroplast and cytoplasmic-type ribosomes. J Cell Sci 107:649–657

McFadden GI, Reith ME, Munholland J, Lang-Unnasch N (1996) Plastid in human parasites. Nature 381:482

McFadden GI, Roos DS (1999) Apicomplexan plastids as drug targets. Trends Microbiol 7:328–333

Mereschkowsky C (1905) Über Natur und Ursprung der Chromatophoren im Pflanzenreiche. Biol. Centralbl. 25:593–604

Molik S, Karnauchov I, Weidlich C, Herrmann RG, Klösgen RB (2001) The Rieske Fe/S protein of the cytochrom b_6/f complex in chloroplasts: missing link in the evolution of protein transport pathways in chloroplasts? J Biol Chem 276:42761–42766

Moore M, Harrison MS, Peterson EC, Henry R (2000) Chloroplast Oxa1p homolog albino3 is required for post-translational integration of the light harvesting chlorophyll-binding protein into thylakoid membranes. J Biol Chem 275:1529–1532

Mori H, Cline K (1998) A signal peptide that directs non-Sec transport in bacteria also directs efficient and exclusive transport on the thylakoid delta pH pathway. J Biol Chem 273:11404–11408

Mori H, Cline K (2001) Post-translational protein translocation into thylakoids by the Sec and pH-dependent pathways, Biochem Biophys Acta 1541:80–90

Mould RM, Robinson C (1991) A proton gradient is required for the transport of two lumenal oxygen-evolving proteins across the thylakoid membrane. J Biol Chem 266:12189–93

Motohashi R, Nagata N, Ito T, Takahashi S, Hobo T, Yoshida S, Shinozaki K (2001) An essential role of a TatC homologue of a DpH-dependent protein transporter in thylakoid membrane formation during chloroplast development in *Arabidopsis thaliana*. Proc Natl Adac Sci USA 98:10499–10504

Muckel E, Soll J (1996) A protein import receptor of chloroplasts is inserted into the outer envelope membrane by a novel pathway. J Biol Chem 271:23864–23852

Nielsen E, Akita M, Davila-Aponte J, Keegstra K (1997) Stable association of chloroplastic precursors with protein translocation complexes that contain proteins from both envelope membranes and a stromal Hsp100 molecular chaperone. EMBO J 16:935–46

Oblong JE, Lamppa GK (1992) Precursor for the light-harvesting chlorophyll a/b-binding protein synthesized in *Escherichia coli* blocks import of the small subunit of ribulose-1,5-bisphosphate carboxylase/oxygenase. J Biol Chem 267:14328–14334

Pain D, Blobel G (1987) Protein import into chloroplasts requires a chloroplast ATPase. Proc Natl Acad Sci USA 84:3288–3292

Palmer JD (2000) A single birth of all plastids? Nature 405:32–33

Payan LA, Cline K (1991) A stromal protein factor maintains the solubility and insertion competence of an imported thylakoid membrane proteins. J Cell Biol 112:603–613

Peltier JB, Friso G, Kalume DE, Roepstorff P, Nilsson F, Adamska I, van Wijk KJ (2000) Proteomics of the chloroplast: systematic identification and targeting analysis of lumenal and peripheral thylakoid proteins. Plant Cell 12:319–341

Perry SE, Buvinger WE, Bennett J, Keegstra K (1991) Synthetic analogues of a transit peptide inhibit binding or translocation of chloroplastic precursor proteins. J Biol Chem 266:11882–11889

Perry SE, Keegstra K (1994) Envelope membrane proteins that interact with chloroplastic precursor proteins. Plant Cell 6:93–105

Pfanner N, Geissler A (2001) Versatility of the mitochondrial protein import machinery. Nat Rev Mol Cell Biol 2:339–349

Pilon M, de Kruijff B, Weisbeek PJ (1992) New insights into the import mechanism of the ferredoxin precursor into chloroplast. J Biol Chem 267:2548–2556

Pilon M, Wienk H, Sips W, de Swaaf M, Talboom I, van't Hof R, de Korte-Kool G, Demel R, Weisbeek P, de Kruijff B (1995) Functional domains of the ferredoxin transit sequence involved in chloroplast import. J Biol Chem 270:3882–3893

Pilgrim ML, van Wijk K-J, Parry DH, Sy DAC, Hoffman NE (1998) Expression of a dominant negative form of cpSRP54 inhibits chloroplast biogenesis in *Arabidopsis*. Plant J 13:177–186

Pinnaduwage P, Bruce BD (1996) In vitro interaction between a chloroplast transit peptide and chloroplast outer envelope lipids is sequence-specific and lipid class-dependent. J Biol Chem 271:32907–32915

Rial DV, Arakaki AK, Ceccarelli (2000) Interaction of the targeting sequence of chloroplast precursors with Hsp70 molecular chaperones. Eur J Biochem 267:6239–6248

Richter S, Lamppa GK (1998) A chloroplast processing enzyme functions as the general stromal processing peptidase. Proc Natl Acad Sci USA 95:7463–7468

Rochaix JD (2001) Posttranscriptional control of chloroplast gene expression. From RNA to photosynthetic complex. Plant Physiol 125:142–144

Rodermel S (2001) Pathways of plastid-to-nuclear signalling. Trends in Plant Science 6:471–478

Roy LM, Barkan A (1998) A SecY homologue is required for the elaboration of the chloroplast thylakoid membrane and for normal chloroplast gene expression. J Cell Biol 141:385–395

Salomon M, Fischer K, Flügge U-I, Soll J (1990) Sequence analysis and protein import studies of an outer chloroplast envelope polypeptide. Proc Natl Acad Sci USA 87:5778–5782

Schein AI, Kissinger JC, Ungar LH (2001) Chloroplast transit peptide prediction: a peek inside the black box. Nucleic Acids Res 29:E82

Schindler C, Hracky R, Soll J (1987) Protein transport in chloroplasts: ATP is prerequisite. Z Naturforsch 42c:103–108

Schleiff E, Tien R, Salomon M, Soll J (2001) Lipid Composition of Outer Leaflet of Chloroplast Outer Envelope Determines Topology of OEP7. Mol Biol Cell 12:4090–4102

Schleiff E, Klösgen RB (2001) Without a little help from 'my' friends: direct insertion of proteins into chloroplast membranes? Biochim Biophys Acta 1541:22–33

Schnell DJ, Kessler F, Blobel G (1994) Isolation of components of the chloroplast protein import machinery. Science 266:1007–1012

Schnell DJ, Blobel G, Keegstra K, Kessler F, Ko K, Soll J (1997) A consensus nomenclature for the protein-import components of the chloroplast envelope. Trend Cell Biol 7:303–304

Schubert M, Petersson UA, Haas BJ, Funk C, Schröder WP, Kieselbach T (2001) Proteome map of the chloroplast lumen of Arabidopsis thaliana. J Biol Chem epub ahead of print

Schuenemann D, Amin P, Hartmann E, Hoffman NE (1999) Chloroplast SecY is complexed to SecE and involved in the translocation of the 33-kDa but not the 23-kDa subunit of the oxygen-evolving complex. J Biol Chem 274:12177–12182

Seedorf M, Waegemann K, Soll J (1995) A constituent of the chloroplast import complex represents a new type GTP-binding protein. Plant J 7:401–411

Small I, Wintz H, Akashi K, Mireau H (1998) Tow birds with one stone: genes that encode products targeted to two or more compartments. Plant Mol Biol 38:265–277

Sohrt K, Soll J (2000) Toc64, a new component of the protein translocon of chloroplasts. J Cell Biol 148:1213–1221

Stahl T, Glockmann C, Soll J, Heins L (1999) Tic40, a new "old" subunit of the chloroplast protein import translocon. J Biol Chem 274:37467–37472

Strub A, Lim JH, Pfanner N, Voos W (2000) The mitochondrial protein import motor. Biol Chem 381:943–949

Sulli C, Schwartzbach SD (1995) The polyprotein precursor to the *Euglena* light-harvesting chlorophyll a/b-binding protein is transported to the Golgi apparatus prior to chloroplast import and polyprotein processing. J Biol Chem 270:13084–13090

Sun Y-J, Forouhar F, Li H-M, Tu S-L, Yeh Y-H, Kao S, Shr H-L, Chou C-C, Chen C, Hsiao C-D (2002) Crystal structure of pea Toc34, a novel GTPase of the chloroplast protein translocon. Nature Struct Biol epub ahead of print

Suzuki S, Izumihara K, Hase T (1991) Plastid import of iron-sulfur cluster assembly of photosynthetic and non-photosynthetic ferredoxin isoproteins in maize. Plant Physiol 97:375–380

Sveshnikov N, Soll J, Schleiff E (2000a) Toc34 is a preprotein receptor regulated by GTP and phosphorylation. Proc Natl Acad Sci USA 97:4973–4978

Sveshnikova N, Grimm R, Soll J, Schleiff E (2000b) Topology studies of the chloroplast protein import channel Toc75. Biol Chem 381:687–693

Theg SM, Bauerle C, Olsen LJ, Selman BR, Keegstra K (1989) Internal ATP is the only energy requirement for the translocation of precursor proteins across chloroplastic membranes. J Biol Chem 264:6730–6736

Theg SM, Geske FJ (1992) Biophysical characterization of a transit peptide directing chloroplast protein import. Biochemistry 31:5053–5060

Tranel PJ, Froehlich J, Goyal A, Keegstra K (1995) A component of the chloroplastic protein import apparatus is targeted to the outer envelope via a novel pathway. EMBO J 14:2436–2446

Tsai L-Y, Tu S-L and Li H-M (1999) Insertion of atToc34 into the chloroplastic outer membrane is assisted by at least two proteinaceous components in the import system. J Biol Chem 274:18735–18740

Tsugeki R, Nishimura M (1993) Interaction of homologues of Hsp70 and Cpn60 with ferredoxin-NADP+ reductase upon its import into chloroplasts. FEBS Lett 320:198–202

Tu CJ, Schuenemann D, Hoffman NE (1999) Chloroplast FtsY, chloroplast signal recognition particle, and GTP are required to reconstitute the soluble phase of light-harvesting chlorophyll protein transport into thylakoid membranes. J Biol Chem 274:27219–27224

Tu S-L, Li HM (2000) Insertion of OEP14 into the outer envelope membrane is mediated by proteinaceous components of chloroplasts. Plant Cell 12:1951–1959

VanderVere PS, Bennett TM, Oblong JE, Lamppa GK (1995) A chloroplast processing enzyme involved in precursor maturation shares a zinc-binding motif with a recently recognized family of metalloendopeptidases. Proc Natl Acad Sci USA 92:7177–7181

van't Hof R, van Klompenburg W, Pilon M, Kozubek A, de Korte-Kool G, Demel RA, Weisbeek PJ, de Kruijff B (1993) The transit sequence mediates the specific interactions of the precusor of ferredoxin with chloroplast envelope membrane lipids. J Biol Chem 268:4037–4042

van't Hof R, de Kruijff (1995) Transit sequence-dependent binding of the chloroplast precursor protein ferredoxin to lipid vesicles and its implications for membrane stabilty. FEBS Lett 361:35–40

Voelker R, Barkan A (1995) Two nuclear mutations disrupt distinct pathways for targeting proteins to the chloroplast thylakoid. EMBO J 14:3905–3914

von Heijne G, Steppuhn J, Herrmann RG (1989) Domain structure of mitochondrial and chloroplast targeting peptides. Eur J Biochem 180:535–545

Waegemann K, Paulsen H, Soll J (1990) Phosphorylation of the transit sequence of chloroplast precursor proteins. J Biol Chem 271:6545–6554

Waegemann K, Soll J (1991) Characterization of the protein import apparatus in isolated outer envelopes of chloroplasts. Plant J 1:149–158

Waegemann K, Soll J (1996) Phosphorylation of the transit sequence of chloroplast precursor proteins. J Biol Chem 271:6545–6554

Waller RF, Reed MB, Cowman AF, McFadden GI (2000) Protein trafficking to the plastid of *Plasmodium falciparum* is via the secretory pathway. EMBO J 19:1794–1802

Westphal S, Heins L, Soll J, Vothknecht UC (2001a) Vipp1 deletion mutant of Synechocystis: a connection between bacterial phage shock and thylakoid biogenesis? Proc Natl Acad Sci USA 98:4243–4248

Westphal S, Soll J, Vothknecht UC (2001b) A vesicle transport system inside chloroplasts. FEBS Lett 506:257–261

Woolhead CA, Thompson SJ, Moore M, Tissier C, Mant A, Rodger A, Henry R, Robinson C (2001) Distinct Albino3-dependent and -independent pathways for thylakoid membrane protein insertion. J Biol Chem 276:40841–40846

Wu C, Ko K (1993) Identification of an uncleavable targeting signal in the 70-kilodalton spinach chloroplast outer envelope membrane protein. J Biol Chem 268:19384–19391

Zimmermann R, Sagstetter M, Lewis MJ, Pelham HR (1988) Seventy-kilodalton heat shock proteins and an additional component from reticulocyte lysate stimulate import of M13 pocoat protein into microsomes. EMBO J. 9:2875–2880

GPSR Compliance
The European Union's (EU) General Product Safety Regulation (GPSR) is a set of rules that requires consumer products to be safe and our obligations to ensure this.

If you have any concerns about our products, you can contact us on

ProductSafety@springernature.com

In case Publisher is established outside the EU, the EU authorized representative is:

Springer Nature Customer Service Center GmbH
Europaplatz 3
69115 Heidelberg, Germany

www.ingramcontent.com/pod-product-compliance
Lightning Source LLC
Chambersburg PA
CBHW071233230926
51849CB00001B/91
* 9 7 8 3 5 4 0 4 3 5 2 0 4 *